Using WinFax to Receive a Fax

WinFax can automatically receive faxes in background, even if another Windows application is running, but if you'r █████████████████ line for both voice and fax, you might not wa █████████████████

To set up WinFax for automatic reception, s █████████████████ dows desktop and select Receive ➤ Automa █████████████████ Receive in the Receive Setup dialog box and n █████████████████ Startup Group, so that it will run automatically when you start Windows.

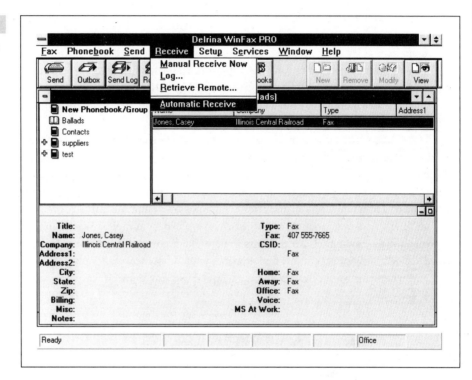

FIGURE 3

The Automatic Receive command in the WinFax Pro 4 Menu

WinFax also includes a memory-resident DOS program for receiving faxes when Windows is not running. If you're using WinFax Lite, use this command at your DOS prompt:

    WINFAX -p<port number> -w<WinFax directory path>

If you're using WinFax Pro, use this command to receive faxes in DOS:

    WFXTSR -w<WinFax directory path>

For every kind of computer user, there is a SYBEX book.

All computer users learn in their own way. Some need straightforward and methodical explanations. Others are just too busy for this approach. But no matter what camp you fall into, SYBEX has a book that can help you get the most out of your computer and computer software while learning at your own pace.

Beginners generally want to start at the beginning. The **ABC's** series, with its step-by-step lessons in plain language, helps you build basic skills quickly. Or you might try our **Quick & Easy** series, the friendly, full-color guide.

The **Mastering** and **Understanding** series will tell you everything you need to know about a subject. They're perfect for intermediate and advanced computer users, yet they don't make the mistake of leaving beginners behind.

If you're a busy person and are already comfortable with computers, you can choose from two SYBEX series—**Up & Running** and **Running Start**. The **Up & Running** series gets you started in just 20 lessons. Or you can get two books in one, a step-by-step tutorial and an alphabetical reference, with our **Running Start** series.

Everyone who uses computer software can also use a computer software reference. SYBEX offers the gamut—from portable **Instant References** to comprehensive **Encyclopedias**, **Desktop References**, and **Bibles**.

SYBEX even offers special titles on subjects that don't neatly fit a category—like **Tips & Tricks**, the **Shareware Treasure Chests**, and a wide range of books for Macintosh computers and software.

SYBEX books are written by authors who are expert in their subjects. In fact, many make their living as professionals, consultants, or teachers in the field of computer software. And their manuscripts are thoroughly reviewed by our technical and editorial staff for accuracy and ease-of-use.

So when you want answers about computers or any popular software package, just help yourself to SYBEX.

For a complete catalog of our publications, please write:

SYBEX Inc.
2021 Challenger Drive
Alameda, CA 94501
Tel: (510) 523-8233/(800) 227-2346 Telex: 336311
Fax: (510) 523-2373

SYBEX is committed to using natural resources wisely to preserve and improve our environment. As a leader in the computer book publishing industry, we are aware that over 40% of America's solid waste is paper. This is why we have been printing the text of books like this one on recycled paper since 1982.

This year our use of recycled paper will result in the saving of more than 15,300 trees. We will lower air pollution effluents by 54,000 pounds, save 6,300,000 gallons of water, and reduce landfill by 2,700 cubic yards.

In choosing a SYBEX book you are not only making a choice for the best in skills and information, you are also choosing to enhance the quality of life for all of us.

This Book Is Only the Beginning.

Just the Fax:
All About WinFax™

JOHN ROSS

San Francisco • Paris • Düsseldorf • Soest

SYBEX®

Developmental Editor: Gary Masters
Editor: Michelle Nance
Technical Editor: Don Rose
Book Designer: Suzanne Albertson
Production Artist: Helen Bruno
Screen Graphics: Cuong Le
Desktop Publishing Specialist: Stephanie Hollier
Proofreader/Production Assistant: Stephen Kullmann
Indexer: Nancy Guenther
Cover Designer: DesignSite
Cover Photographer: Mark Johann

ACKNOWLEDGMENTS

THANKS are due to Shelly Sofer, Dennis Tapaluca, and Rob Stewart at Delrina for their advice, assistance, and guidance into the details of WinFax. At SYBEX, this book has been greatly improved by the editorial efforts of Michelle Nance, by Don Rose's technical review, and by all of the others responsible for design, layout, art, production, and distribution. A special thank you goes to Developmental Editor Gary Masters for assigning this project to me in the first place, and for his continuing aid and counsel through the process of creating the manuscript.

John Ross

Contents

AT A GLANCE

CONTENTS

PART III   WINFAX PRO

INTRODUCTION

DELRINA'S WinFax fax modem control software is the best-selling product for sending and receiving faxes through a PC. More than two million copies of the program have been sold. If you're one of those millions of users, or if you're looking for information about getting started using a fax modem, this book contains the information you need to get the most out of WinFax.

If you're like most WinFax users, you probably installed the program to send and receive faxes through your PC, just as you would use a stand-alone fax machine. But some WinFax features allow you to do things you can't do on a stand-alone fax, such as receiving a fax and copying its text and graphics directly to a spreadsheet, database, or word processor document. Once you've mastered the WinFax features explained in this book, you'll discover that WinFax is an even more flexible tool than a conventional fax machine.

Whether you obtained a copy of WinFax Lite with your fax modem or bought WinFax Pro separately, this book will help you install WinFax, tell you how to send and receive faxes, and introduce you to the program's advanced features. If you're using WinFax Lite or Version 3 of WinFax Pro, you can also find information in this book that will help you decide whether to upgrade to WinFax Pro 4.0.

Hardware and Software Requirements

To use WinFax, you must have a PC with Windows 3.0 or later and a fax modem. If Windows works on your PC, you almost certainly have enough memory for the WinFax features, with the possible exception of the Recognize (optical character recognition) feature of WinFax Pro, which requires at least 4.5 megabytes of RAM. However, if your PC has fewer than 4 megabytes of RAM, you should seriously consider investing in additional memory. WinFax and just about every other Windows application will run more efficiently when you install more RAM.

WinFax won't do you any good unless you have a fax modem. Most new modems for PCs support both data and fax transmission, but some older modems only work with data. Before you try to use WinFax, make sure your modem recognizes fax signals. Look for the word "fax" on the front or rear panel of the modem, or in the modem manual.

If you don't already have a fax modem, look for one that can operate at 14.4 kbps or faster for both fax and data transmission. WinFax will work with slower modems, but prices are dropping so quickly that you might as well get a high speed model. Since you pay for long-distance calls by the minute, a faster modem will pay for itself in reduced telephone bills.

WinFax supports most makes and models of fax modems, including those made by all major manufacturers such as Hayes, U.S. Robotics, Supra, and Intel. If your modem is compatible with Class 1, Class 2, or CAS software, it should work with WinFax.

And of course, you need a telephone line for your modem. You can use the same telephone number that you use for voice, or you can dedicate a separate line for fax and data calls. You can find information about sharing a single line for voice and fax in Chapter 1.

Versions of WinFax

This book describes three different versions of WinFax: WinFax Lite, WinFax Pro 3.0, and WinFax Pro 4.0. You can find information about using WinFax Lite in Part II, and information about both versions of WinFax Pro in Part III

WinFax Lite is not sold directly to end users, but many manufacturers bundle it with their modems. WinFax Lite is very similar to Version 2.0 of WinFax Pro, which is no longer available. It includes all the functions necessary to send faxes from Windows applications and to receive incoming faxes whenever your computer is turned on, even if Windows is not currently running.

For many users, WinFax Lite is the only fax software you'll need. However, Delrina has introduced new features and functions in each new version of WinFax. In Version 3.0 of WinFax Pro, you can make annotations to received faxes before you print or re-send them, convert and export the text of a fax to another application, and eliminate flyspecks and other interference from a fax before you print or re-send it. If you have a TWAIN-compatible scanner, you can also scan paper documents directly into WinFax Pro.

Version 4.0 of WinFax Pro was released early in 1994. It includes several significant improvements over the earlier versions of the product. Among other things, the WinFax screen more closely resembles other Windows applications, and it's no longer necessary to convert files to a special attachment format before you include them in a fax. Since fax modems are becoming more common, WinFax Pro 4.0 can transfer documents to another fax modem without first converting them to fax images, which improves both transmission speed and image quality. And there are new autoforwarding, polling, and error correction mode functions that were not available in the earlier versions.

How This Book Is Organized

Part I of this book contains general information about using a modem and about sending and receiving faxes. You will find this information useful with any version of WinFax. Part I also contains step-by-step procedures for configuring your computer, Windows, and your modem to work together.

Part II is a detailed description of WinFax Lite. If you want a guide to the WinFax software that came with your modem, this is the place to look for information about installing the program and using it to send and receive faxes.

Part III describes Versions 3.0 and 4.0 of WinFax Pro. Read this section to learn about the features of the shrink-wrapped WinFax product, including the advanced functions for annotation, text recognition, and cover page creation. If you're considering upgrading from WinFax Lite, or from Version 3.0 to Version 4.0, you can evaluate these added features before you commit yourself to spending more money.

Part IV contains troubleshooting information for all versions of WinFax. When WinFax doesn't perform the way you expect it to, or if the program displays cryptic messages on your screen, look in this section for quick advice.

Finally, Appendix A covers the WinFax specifications for Dynamic Data Exchange, and Appendix B explains variations between the different versions of WinFax Lite.

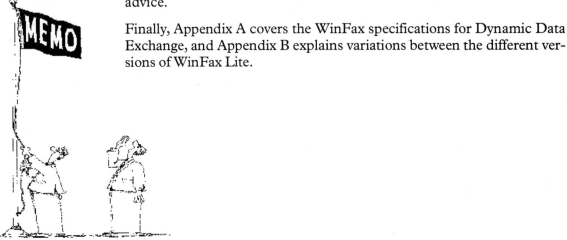

Introduction to WinFax

WinFax is a great tool for sending and receiving faxes through your computer, but it doesn't work all by itself. In this section, we'll give you some background on faxing from your PC and tell you how to make sure your PC, your modem, and Windows are all operating properly. Finally, we'll explain the differences between WinFax Lite and WinFax Pro.

CHAPTER

1

An Overview
of WinFax

EVERY day, millions of fax messages move across telephone lines, carrying everything from multimillion dollar contracts and marriage proposals to lunchtime sandwich orders. Over the last few years, the fax machine has become a universal office tool. It's just about impossible to run any kind of business without one. But when you use a desktop computer to write letters, reports, and memos, and to create graphic images and design documents, it can be a time-consuming nuisance to print something onto a piece of paper, just so you can feed it to your fax machine.

Delrina's WinFax family of Windows applications can send and receive fax messages through your desktop computer, without the need for a separate fax machine. In exactly the same way a Windows printer driver converts text and images in a computer file to instructions your printer will recognize, WinFax converts the same text and images to facsimile signals. Using an inexpensive accessory called a *modem*, you can send a copy of a document created with a word processor, spreadsheet, or any other Windows program to a distant fax machine (or another computer with a fax modem) as easily as you send it to the printer on your own desk. Win-Fax can also receive incoming fax messages in the background, while the computer is running other programs. It even includes a memory-resident DOS program to receive new faxes when Windows is not running.

Many fax-compatible modems come with the entry-level WinFax Lite fax program that sends and receives fax messages, and allows you to export received fax images to other applications. The more sophisticated WinFax Pro package is sold separately through software retailers, and with several other office productivity applications in bargain combination packages. If you're a registered WinFax Lite owner, you can upgrade to WinFax Pro for about $50, which is substantially less than the best street price for the same product.

The Pro version includes all the functions of WinFax Lite, and adds many more advanced features. In the newest release of WinFax Pro, Version 4.0, you can use binary file tranfer to attach data files to your faxes, when the recipient is able to accept them.

This book explains how to use both WinFax Lite and WinFax Pro. This chapter explains how facsimile works, and provides some general information about sending and receiving faxes on a computer. It also includes some suggestions for using the same telephone line for both voice and fax, and describes the differences between WinFax Lite and WinFax Pro, and between Versions 3.0 and 4.0 of WinFax Pro.

Part II of this book contains everything you need to know to install WinFax Lite, and to use that program to send and receive fax messages. Part III is a complete guide to WinFax Pro, including all of its advanced features. Part IV contains troubleshooting information for both WinFax versions. You can find a complete WinFax Lite command summary in the inside front cover. The WinFax Pro command summary is inside the back cover.

How a Fax Works

If you're used to sending and receiving faxes through a stand-alone fax machine, you probably think of the process as inserting a piece of paper into one machine and having an exact copy—a *facsimile*—appear on a second machine in the next office, or halfway around the world. That's essentially accurate, but there are actually several parts to the process. First, the sending machine makes a telephone call to another fax machine. When the connection is in place, the originating machine uses a spot of light to scan the paper fed into it, converts those images into electrical impulses, and sends the image to the distant machine as a series of sounds. Finally, the receiving machine converts those sounds back into electrical impulses, which it uses to print a copy of the original document.

As long as the signal format meets the international standards for fax messages, it's easy enough to replace the scanner at one end, or the printer at the other end, with a computer that recognizes fax signals and converts between those signals and graphic images, which it can store as a file, display on a monitor, or print. If both ends speak the same electronic "language," you can send a fax message from a stand-alone fax machine to a computer, from a computer to a fax machine, or from one computer to another.

Fax Machines vs. Modems: Which Is Better?

Both fax machines and modems have their advantages and disadvantages. You can leave a stand-alone machine on all the time and not worry about missing an incoming message. Everybody in your house or office can walk up to the machine, push a few buttons, and immediately send a fax. And when a new fax message arrives, it's easy to carry it to the intended recipient.

On the other hand, many fax machines use special rolls of slimy, expensive paper that's unpleasant to handle. If you leave a fax in the light or on top of a warm surface, it turns brown and fades. Unless you place the original in the machine exactly right (and sometimes even then), the copy at the other end is full of jaggies and flyspecks. And if the company fax machine isn't in your own office, you've got to interrupt what you're doing and walk across the office or down the hall to send and receive your faxes. When a new fax message arrives, anybody who walks past the machine can read it; sharing a fax machine is not a good way to keep your messages private.

But faxing through a computer isn't perfect either. It can be convenient to send and receive faxes without leaving your desk, but you have to leave the computer on all the time to make sure you don't miss an incoming fax. You can print copies of faxes on ordinary paper, and make as many extra copies as you need. But since the computer treats fax images as graphics, each page takes something just short of forever to print. And you can't send a fax of a newspaper clipping or hand-drawn picture without a separate scanner.

So which is the better way to fax, machine or modem? Like much of life, it depends on what you want to do. For sending memos and reports from your desktop, a fax modem can save a bunch of time and trouble. But when you want to check the boxes on the order blank for pepperoni and extra cheese and send it back to Luigi's Pizza by fax, you're out of luck without a stand-alone machine. Of course, printing out that memo and feeding it into the machine isn't *that* much trouble, and a note that says, "Yo, Luigi! Please bring a large No. 5 with extra sausage to 238 Elm Street" will probably get you your pizza just as quickly. With a few exceptions, you can generally get by with either method.

Since you're reading this book, you're probably most interested in learning what you can do with a fax modem. But in a random survey of WinFax users, Delrina discovered that more than 60 percent also have a stand-alone fax machine. If you're in that group, it's worth remembering that nobody's forcing you to use WinFax for absolutely everything. If it's more convenient to send through WinFax and receive through a separate machine, that's fine—go ahead and do it that way. Remember, your real goal is exchanging messages with another person, not using a computer or fax machine.

The Difference between Faxes and Data

N O T E   This section contains a somewhat technical explanation of the way faxes and computer data move through telephone lines. If terms like *bit, byte,* and *protocol* make your eyes glaze over, you can skip this part. It won't be on the final exam.

Both facsimile images and computer data can move through communications circuits (such as telephone lines). The techniques are similar, but they're not identical. You could compare this to the difference between a toaster and a waffle iron. You use both to heat up breakfast foods, but no matter how hard you try, you can't make waffles by pouring batter into your toaster. In the same way, you can't call a computer bulletin board and download a text or graphics file directly to your fax machine.

Facsimile and computer data both use a system called *serial communication,* which moves one binary bit at a time from the sender to the receiver. In other words, the communications circuit is in one of two states for a specific period of time (sometimes called 1 and 0, or plus and minus, or high and low). Each one or zero is called a *bit.* The sender uses a standard timing arrangement (called a *protocol*) to organize the ones and zeroes in a particular order. When those ones and zeroes arrive at the other end of the communications link, the receiver uses the same protocol to convert them back into information.

Fax images and computer data start as different kinds of information, and they use different protocols to organize them for serial communication. A fax machine scans the page at a rate of up to about 200 lines per inch, looking for tiny light and dark areas. It converts each of those light and dark areas to a piece of binary data. At the other end of the link, the fax machine converts the binary data back into light and dark areas on a page.

Computer data is quite different. Text and data are organized into eight-bit bytes, each of which represents either an individual letter, number, or other text character, or an instruction to the computer. When two computers establish a communications link, they exchange those eight-bit bytes, which both recognize as characters or instructions.

In order to convert binary information to sounds that can move through a telephone line, you need a device called a *modulator*. To convert back to data at the other end, you need a *demodulator*. Since data moves in both directions, a single device combines them into a single box or circuit board called a MOdulator/DEModulator, better known as a *modem*. Modems are available both as expansion boards that fit inside a PC and as stand-alone devices that communicate with a PC through a serial communications port.

A fax modem is a modem that produces and recognizes facsimile signals from a computer. It receives a description of each page from a program like WinFax and converts it to the standard facsimile protocol, which it sends through a telephone line. At the other end, it converts the data back to individual page descriptions, which WinFax (or similar software) displays or prints. Like a human interpreter who reads a page of Greek and translates it to English, a fax modem receives information intended for a fax machine and translates it to computer data.

Many modems can recognize both data and fax protocols, but they won't transform a signal from one to the other. When a dual purpose modem receives a call, it identifies the type of signal it contains and uses the appropriate protocol to convert the incoming signal to either data or fax.

When you start using binary file transfer (BFT), the line between faxes and data becomes fuzzy. In WinFax Pro 4.0, you can attach a binary data file to a fax message, and send a copy to someone else who has a BFT-capable fax modem and software. Unlike a conventional fax image, WinFax does not convert a BFT attachment to fax image format.

Using the Same Telephone Line for Voice and Fax

It would be nice to have separate telephone numbers for your telephone and your fax modem (or stand-alone fax machine), but it's not always practical. Especially if you're sending and receiving faxes from home, the cost of an extra line may be greater than any possible benefit it provides.

When you share a single line for voice and fax, you have several options for handling incoming calls. The "low-tech" approach is to pick up the handset when the telephone rings, say "hello" or something similar, and listen. If somebody starts talking, you can start your conversation. But if you hear a high-pitched tone, it's probably a fax call. Start up the fax machine (choose Manual Answer in WinFax), and let it start receiving. Not very complicated, but difficult if you take the call on your kitchen extension and the computer or fax machine is in the spare bedroom. And when you're not home, your answering machine won't transfer the call to the fax machine. If you set the fax modem to answer automatically, human callers will get an earful of high-pitched tones when they try to call you.

There's a simple solution to this problem. A *line-sharing device* is a telephone accessory that answers incoming calls and listens for the distinctive fax calling tone (known to telephone types as a CNG tone). If there's a fax signal, the line-sharing unit automatically switches the call to the fax machine or modem. If it's a person calling, the line-sharing device doesn't hear the CNG tone, so it connects the call to a telephone instrument and makes the telephone start ringing. The whole process takes a few seconds, and your callers never know the difference. Some line-sharing devices have additional features that direct calls to separate fax machines and modems, but for use with WinFax and a fax modem, you can get along with a basic model that should cost less than $100.

At least a dozen companies make line-sharing devices, including Command Communications, Crest Industries, and Viking. AT&T also makes a line-sharing device, but it costs much more than other brands that work just as well. You can find line-sharing devices at retail telephone stores and large office supply dealers.

WinFax Lite and WinFax Pro: What's the Difference?

Delrina makes two versions of its WinFax product: WinFax Lite and Win-Fax Pro. As we mentioned earlier in this chapter, WinFax Lite is not sold separately, but it's packaged with many fax modems. WinFax Lite contains everything you need to send and receive faxes through your modem. You can send faxes directly from Windows applications, receive in background, and create phonebooks that list frequently-called numbers. Win-Fax Lite uses the Windows clipboard to export graphic images from fax messages to other Windows applications.

WinFax Lite is similar to version 2.0 of WinFax Pro. WinFax Pro 3.0 was released in 1992, and it includes many additional features that were requested by users. The basic design philosophy is the same, but the user interface is quite different.

WinFax Pro 3.0 includes these features that are not in WinFax Lite:

- Optical Character Recognition (OCR) to convert faxes to editable text
- Real-time viewing of incoming faxes
- Optional assignment of billing codes to outgoing fax calls
- Use of dBASE files as phonebooks
- More detailed information in phonebooks
- Clean-up on received faxes of random black spots caused by noisy telephone lines
- Improved on-screen appearance
- Support for scanners
- Improved fax management and archiving
- Mark-up and drawing tools for annotating faxes
- Improved cover sheet designer

- Library of more than a hundred cartoon cover sheets
- Optional reduced-size (four per page) printing
- "Fax button" macros for Word for Windows, Ami Pro, and Excel.

Should you upgrade? WinFax Lite is a complete, self-contained package that can meet your basic fax requirements, but WinFax Pro gives you a lot of extra flexibility. The rest of this book describes all the features in both products; if you're already a WinFax Lite user, take a close look at the descriptions of WinFax Pro. If any of those added features match the way you work—or the way you'd like to work—go ahead and upgrade. But if you're satisfied using Lite to send and receive faxes and you don't expect to need scanning, OCR, annotation, or the other bells and whistles in Pro, there may not be any need to change.

What's New in WinFax 4.0?

Delrina released Version 4.0 of WinFax Pro early in 1994. WinFax Pro 4.0 has a different user interface and several new features: there's a new Quick Cover Page that you can use instead of the graphics in the Cover Page Library; you can add files to a fax without converting them to attachment format first; WinFax can automatically forward received faxes; and you can store more than one standard dial prefix if you use the same computer in different locations. In addition, WinFax Pro 4.0 supports several new fax modem features, including binary file transfer, two-way polling, and error correction mode.

In order to make this book as useful as possible for owners of WinFax Pro, we've included information about both versions 3.0 and 4.0. Where there are differences between the two, Part III includes pictures and descriptions of both. Detailed descriptions of the new features in Version 4.0 are in Chapter 16.

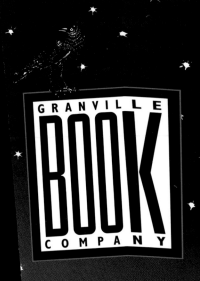

GRANVILLE
BOOK
COMPANY

Open 7 days
& 7 nights

850 Granville Street
Vancouver, BC
Phone 687-2213
Fax 669-3556

CHAPTER

2

Configuring Your System for WinFax

A stand-alone fax machine has one definite advantage over a computer with a fax modem and software: it's easy to install. You can take it out of the box, plug in the telephone line and the power cord, and start sending and receiving faxes within about a minute. (Maybe two minutes if you need to find an extension cord.)

Installing a fax modem and telling Windows how to communicate through it can be a lot more complicated. Before you can use WinFax, you've got to connect the modem to your computer (or install it in an expansion slot if it's an internal modem), make sure the modem doesn't conflict with anything else, and set up Windows to recognize your modem. You could spend hours trying to get all this stuff to work together.

Making your modem work with Windows and WinFax isn't difficult, but you frequently need to pull information together from several places before you start. In this chapter, you can find all the information in one place that you'll need to set up your modem. It contains a very general explanation of computer communication and instructions for testing your modem setup. If you can't complete the test successfully, I'll explain how to make the necessary changes to your modem and your Windows configuration.

If you're already using a Windows data communications program, such as the Windows versions of ProComm Plus, Crosstalk, or SmartCom, you can be pretty certain you've got everything configured properly. If that's the case, you can skip the rest of this chapter. But if you haven't been able to get Windows to talk to your modem, the "Setting Up the Modem to Work with WinFax" section should contain the information you need to get started.

Understanding I/O Ports

This section contains background information about your computer's serial communication ports. The next section is a "cookbook" for testing and configuring your modem for WinFax.

Your computer uses several different interfaces to communicate with the rest of the world. It receives instructions and data from a keyboard and sends information out to a video monitor, it reads and writes data to disk drives, and it uses at least two types of general purpose input/output (I/O) ports, called *parallel* and *serial*. You can use an I/O port to communicate with a printer, another computer, or a communications device such as a remote terminal or a fax machine.

The difference between parallel and serial ports has to do with the way data moves in and out of the computer. A parallel communications link uses eight separate wires to move eight bits of data at the same time. In a serial link, the data moves one bit at a time through the same wire, and the device that receives the data assembles those bits into eight-bit characters and instructions. Because more information is moving at one time, you can send data through a parallel link more quickly, but the cost is greater.

In practice, the difference in cost is insignificant over very short distances, so most PCs use a parallel port to send data to a printer. But when you're exchanging data with another computer several miles (or several thousand miles) away, it's much more practical to use a serial connection. The designers of your PC recognized this fact, and that's why they labeled the parallel connectors as "printer ports" (DOS calls the ports LPT*n*, which stands for "line printer"), and the serial connectors as "COM (short for communication) ports." As explained in Chapter 1, both computer data and faxes are forms of serial data, and therefore you must connect your modem to one of your computer's serial I/O ports. DOS recognizes each I/O port as a unique port address.

Avoiding the Dreaded Interrupt Conflict

One more complication arises when you try to use an internal modem. Even though your computer probably has only one or two built-in serial

ports (COM1 and COM2), you can set switches or jumpers to configure the modem with the port address for COM1, COM2, COM3, or COM4.

If a computer spent all of its time exchanging data through a communications port, there would be a direct link from the I/O port to the computer's processor. But real life is not that simple. As far as your PC is concerned, a request for attention from a communications port interrupts whatever other work the computer might be doing at the time. Your PC uses specific interrupt requests (IRQs) as signals that a communications port requires attention from the CPU. Normally, DOS uses IRQ4 for serial ports COM1 and COM3, and IRQ3 for COM2 and COM4. The fact that the same IRQ number controls two COM ports can create problems when you need to use more than one serial port at a time; IRQ conflicts are one of those fiddly configuration difficulties that make installing peripheral devices in a PC such an adventure.

If two devices try to use the same IRQ, don't expect either one to work reliably. When you install an internal modem, you must select a base address (serial port number) with an IRQ that doesn't conflict with a device that's already in place. In practice, this means you should assign the modem to COM4 (Base Address 02E8) if you have a mouse or other device plugged into COM1, or assign the modem to COM3 (03E8) if you're already using COM2.

If you're already using both COM1 and COM2, things get a little messier, because you can't completely avoid a conflict. The best you can do is to share the same IRQ between two devices that you don't expect to use at the same time.

Setting Up the Modem to Work with WinFax

In most cases, installing your modem is relatively simple—connect the cables to a serial port on the back of your computer or insert the internal fax/modem board in an empty expansion slot. If everything is set to the default configuration, you should be able to start sending and receiving data and fax messages right away.

Here's how to test your installation:

1. Start the computer and turn on the modem's power switch. If you're using an internal modem, it will turn on automatically when you turn on the computer.

2. Start Windows. If you've already installed WinFax, make sure it is *not* running.

3. Open the Windows Terminal Program. The Terminal icon is in the Accessories group.

4. Select Settings ➤ Communications. Figure 2.1 shows the Communications dialog box.

FIGURE 2.1

The Terminal
Communications
dialog box

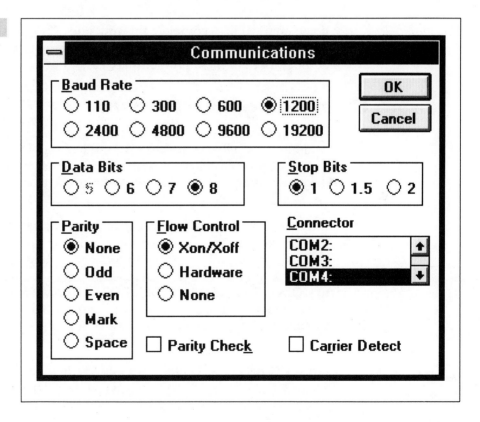

5. In the Connector box in the lower right side of the dialog box, highlight the COM port connected to your modem. For this test, you can ignore the other Communications settings. Click on OK. If Windows gives you an error message, close the Terminal program and complete the configuration routines in the next two sections of this chapter.

6. If the Terminal program accepts your setting, type **ATE1** and press ↵.

7. If the installation is correct, you will see an *OK* message from your modem. If that happens, your modem is configured properly and you can skip the next two sections of this chapter. If you're using WinFax Lite, move ahead to Chapter 3. If you're using WinFax Pro, jump to Chapter 8.

If you aren't able to complete this test, one of two things is wrong: you're using either the wrong COM port or the wrong interrupt request (IRQ) number. Both of these problems are easy to fix, as you'll see below.

Configuring Your Modem

External modems come in their own housing and are connected to the computer through a cable. Internal modems are located in an expansion slot inside your computer. Installing an external modem is easy: connect the power, plug in the cable, connect it to the telephone line, and turn it on. Simple enough.

Internal modems are a little more complicated. If you're using an internal modem, you must set jumpers or switches to specify which serial port your computer will use to communicate through the modem. If your computer has only one built-in serial port, you should assign the modem to COM2 (Base Address 02F8, IRQ3). If COM2 is already assigned, you can define the internal modem as either COM3 or COM4, depending on which is least likely to create an interrupt conflict.

The manual supplied with the modem should explain how to set the interrupt request (IRQ) and base address jumpers or switches. WinFax works with a long list of modem types, and each manufacturer has its own configuration method. Follow the installation instructions supplied with your modem.

Remember which COM port you've assigned to the modem; you may need that information when you install WinFax. If you're using an internal modem, you may also need to tell Windows where to find the modem.

Configuring Windows for an Internal Modem

If you've installed your modem on serial port COM3 or COM4, you'll need to make some changes to your Windows configuration. There is one procedure for Windows 3.0 and another for Windows 3.1.

In Windows 3.0, use a text editor to add these lines to the [386Enh] section of the SYSTEM.INI file:

```
COM3BASE=03E8H
COM4BASE=02E8H
COM3IRQ=4
COM4IRQ=3
```

If you're using an IBM PS/2, use these lines:

```
COM3BASE=03E8H
COM4BASE=02E0H
COM3IRQ=4
COM4IRQ=3
```

In Windows 3.1, run the Windows Control Panel program (in the Main Group), and double click on Ports. When the Ports window appears, select the port that you're using for your modem. Click on Settings and then on Advanced. Figure 2.2 shows the Advanced Port Settings dialog box.

FIGURE 2.2

The Windows Advanced Port Settings dialog box

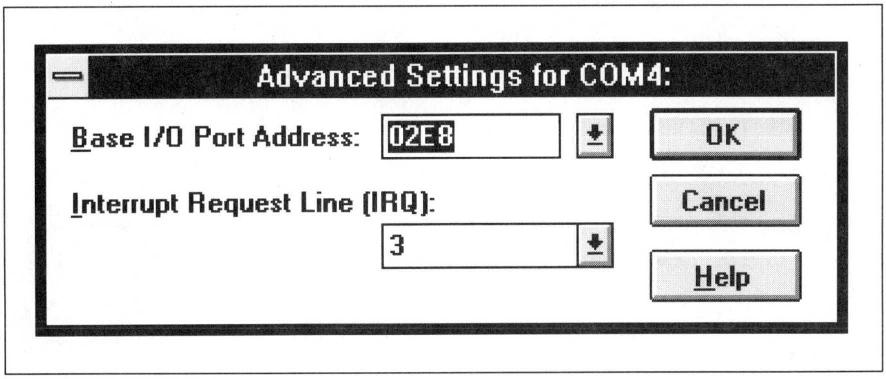

If your modem is on COM3, set the Base I/O Port Address to 03E8, and set the Interrupt Request Line (IRQ) to 4. If your modem is on COM4, set the Base I/O Port Address to 02E8 and the Interrupt Request Line (IRQ) to 3.

If you have an IBM PS/2 computer, the Base I/O Port Address for COM4 is slightly different. If you're installing a modem in serial port COM4 on a PS/2, the Port Address should be 02E0.

Before you can send and receive faxes with WinFax, you must install a fax modem and make sure Windows knows where to find it. If you've been able to complete the test in this chapter, you should be able to install WinFax without much pain. If you're using WinFax Lite, you can find complete information about sending, receiving, and managing faxes in Part II of this book. If you're a WinFax Pro user, you'll find lots of useful information in Part III. If you haven't installed the program yet, look at Chapter 3 (Installing WinFax Lite) or Chapter 8 (Installing WinFax Pro) for step-by-step instructions.

WinFax Lite

Many modem manufacturers include a copy of WinFax Lite with every fax modem they sell. WinFax Lite has fewer advanced features than the more costly WinFax Pro, but it is a solid fax modem control program. In this section, we'll give you all the information you need to install WinFax Lite, and we'll tell you how to use it for sending, receiving, and managing fax messages.

WinFax Lite and WinFax Pro are two quite different programs. If you're looking for information about WinFax Pro, you should skip Part II and jump immediately to Part III.

CHAPTER

3

Installing
WinFax Lite

IN the best of all possible worlds, you could install a Windows application by just sticking the distribution diskette into your drive and typing **INSTALL** or **SETUP**. Unfortunately, WinFax Lite requires a little more attention before you can start sending and receiving faxes. In addition to loading the software, you must also configure Windows to recognize your modem and the WinFax program. This chapter contains complete instructions for all the things you need to do in order to get WinFax Lite up and running.

Before you install the WinFax program, you must install Windows 3.0 or 3.1 on your computer, and either attach an external modem to one of the computer's COM ports or install an internal modem in an expansion slot inside the computer. You can use the same modem to send and receive both data and fax messages, but you'll need a separate data communications program to move data, such as ProComm Plus, SmartCom, Crosstalk, or the Windows Terminal program.

Loading and Configuring WinFax

After you've installed the modem, you're ready to load WinFax. Here's the procedure:

1. If it's not already running, turn on your computer and start Windows.

2. Place the WinFax diskette into a drive, use the Run command in the Program Manager menu bar, and enter this command:

   A:\INSTALL

   Use the correct drive letter for the drive that contains the WinFax diskette.

3. The program displays a sequence of dialog boxes; select OK on each to move to the next until the Printer Driver Setup dialog box appears, as shown in Figure 3.1.

FIGURE 3.1

The Printer Driver
Setup dialog box

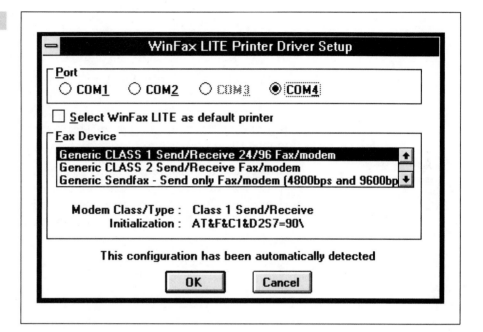

4. The Printer Driver Setup dialog box shows the port number and modem type detected by the setup program. Unfortunately, the program isn't always able to find your modem. If the dialog box shows the wrong port number or modem type, click on the correct COM number. If the COM port assigned to your internal modem is grayed out, you won't be able to select it. If this happens, click on Cancel to stop the WinFax Setup program, and follow the procedure for adding the modem port in the "Configuring Windows for an Internal Modem" section in Chapter 2. After you've completed that process, run the INSTALL program again.

5. Use the up and down arrows to the right of the list of fax devices in the Driver Setup dialog box until a description of your modem appears. Highlight that device type. If you can't find an exact match, select one of the Generic devices.

The Fax Device list identifies some fax devices as *CLASS 1*, *CLASS 2*, or *Sendfax*. Your modem manual should identify the correct protocol for your modem. Class 1 and Class 2 are standard software protocols for communication between a computer and a fax modem. In Class 1, the computer controls the communications session, while in Class 2, control takes place in the modem, which reduces the amount of work the computer has to perform. If you're running WinFax on a slow PC or one with limited memory, Class 2 may make fewer demands on overall system resources.

6. The Printer Driver Setup dialog box also offers the option of making WinFax Lite your default printer. Unless you expect to send more documents as fax messages than you expect to print, ignore this option. It's simple enough to select WinFax from an application when you're ready to send a fax message.

7. At this point, the program is ready to copy files from the diskette to your hard drive. Either accept the default C:\WINFAX directory or specify a new directory name, and click on OK. The program will display its progress as it writes the files.

8. When the program asks for a path for phonebooks and status files, go ahead and accept the default unless you have a reason to store this information under a different directory name.

9. When the Setup dialog box appears, select the configuration items you want WinFax to use when it transmits fax messages. The configuration of incoming messages will be defined by the fax machine that originated the message.

The Setup dialog box includes these items:

Paper Size   This option specifies the dimensions of the paper your fax messages will be printed on. Click on the arrow at the right of the listing to select either Letter ($8^1/_2 \times 11$ inches), Legal ($8^1/_2 \times 14$ inches), or A4 (210×297 mm).

Default Resolution   This option specifies the number of dots per inch that the distant fax machine will use to print the transmitted image. In most cases, you should select High Resolution for the best possible image quality. But if you want to hold down your long distance telephone charges, use Low Resolution to reduce the amount of time needed to transmit each page.

pencil me in?

I'd like to meet with you!

Orientation   WinFax offers two orientation options—portrait and landscape.

Display Call Progress   This option instructs WinFax to display a dialog box that shows the current status of the transmission when WinFax is transmitting a fax message.

The More button opens the Fax/Modem Setup dialog box, which includes configuration items that you will probably set once and not change. It includes these items:

Modem Setup   Type identifies the modem you specified in the Printer Driver Setup dialog box.

The Initialization is the modem initialization command string that WinFax will send the modem at the start of each transmission. WinFax automatically chooses a command string that's appropriate for your modem.

You can change the initialization string, but the default is probably the best place to start. Your modem manual contains a description of the commands your modem will recognize. If you send fax messages to foreign destinations, you may want to add this command to the command string:

ATS7=120

This command increases the number of seconds the modem will wait to determine whether there's another fax machine at the other end of the call.

Speaker Mode   This option specifies when the modem's speaker will be active. Choose Until Connection to listen to the modem as it establishes the connection. (You cannot listen during the time it transmits or receives a fax message.)

Dial Setup   Make Detect Busy Tone active to automatically terminate the call when the called number is busy.

Make Detect Dial Tone active to wait for a dial tone before the modem starts calling a number.

Make Pulse Dial active if your telephone line requires dial telephones instead of pushbutton tone telephones. Don't select this function if you have a Touch-Tone™ or similar phone system.

Set the number of Dial Retries to the number of times you want WinFax to attempt to make a connection if the called number is busy. Retry Time specifies the number of seconds between retries.

Use the Dial Prefix to add a number string to the beginning of every outgoing call. If your telephone system requires that you dial a *9* or some other number for an outside line, you can use the Dial Prefix to add that number to every call.

Max Tx Rate   This option sets the maximum transmission rate. Select the highest speed your modem can handle; if you're not sure, look in your modem manual.

Volume   This option sets the loudness of the modem's speaker. If Speaker Mode is set to Off, the volume setting doesn't make any difference. Don't use High volume if you have the speaker option set to On rather than Until Connection, because some modems' speakers can overload from excessive volume level.

Fax Header   This option specifies the information that WinFax places on each page it transmits. In the United States, federal law requires that you include the date and time of the transmission and an identification and telephone number of the person or business sending the message on the top or bottom margin of every page, or on the first page of every transmission. WinFax accepts these variables in a fax header:

$D	Current date
$T	Current time
$P	Current page number
$N	Total number of pages
$S	Sender
$R	Recipient name
$X	Recipient telephone number

So if you specify this fax header:

Left	From $S
Center	DT
Right	Page $P of $N

the top of your fax messages will look like this:

From The Fax Message Center   11/4/93 12:12:04   Page 1 of 1

Station Identifier   The Station Identifier is the information that WinFax returns to the distant fax machine to identify your line. Many fax machines and programs (including WinFax) use the Station Identifier in fax activity lists or digital readouts. Most people use their telephone number as their Station Identifier.

Sender Name   Type the name you want WinFax to use in fax headers. In the example above, this option would contain the words *The Fax Message Center*.

When you have answered all of the questions in the Setup dialog box, click on OK to complete the installation. WinFax will offer you the chance to test your installation by sending a fax message to Delrina registering your software. Fill in the blank spaces and click on Send.

If you ever need to change any of your settings, you can display the same Setup dialog boxes when you're ready to send a fax message. It's also possible to make changes from the WinFax Lite Administrator program, or directly to the [WINFAX] section of your WIN.INI file.

The [WINFAX] section of WIN.INI looks something like this:

```
[WINFAX]
ConnCount=3
ModemDesc=Supra Corp. - Supra FAXModem V.32bis Internal
modem=AT&F&C1&D2&K4S95=44S7=90\
Fax Device=Class1
Fax Path=C:\WINFAX\data\
ExePath=C:\WINFAX\
Paper Format=Letter (8.5 x 11 inches)
Orientation=Portrait
Dial Prefix=
Resolution=Standard
Retries=0
Call Progress=Yes
CSID=206-555-8562
Sender=Ballard Ballad Works
Header Left=From $S
Header Center=Fax: 206-555-8562
```

```
Header Right=$D $T  Page $P of $N
Max Tx Rate=9600
Retry Time=60
Dial Control=All
Volume=Med
Speaker Mode=Connect
Dial Mode=Tone
```

If you add this command:

```
Priority=High
```

WinFax disables background operation, which means you won't be able to do anything else while WinFax is transmitting a fax message. This doesn't sound especially practical, but it is possible. To permit background operation, either add this command:

```
Priority=Low
```

or leave the Priority= command out completely.

Registering Your Software

Delrina is pretty insistent about making you register WinFax Lite. They make it relatively painless, providing an automatic fax upload and a toll-free telephone number. If you don't send them your registration during installation, the registration screen will appear every time you try to use the program.

Unfortunately, the automatic registration-by-fax system at the other end of that toll-free number doesn't always accept your upload. If that happens, the program forces you to keep trying to register. Eventually, Win-Fax displays a toll-free voice telephone number that allows you to register without sending a fax. When you call, the operator will give you a code that instructs the program to stop displaying the registration screen.

In this chapter, we explained how to load and configure WinFax Lite. After you complete the installation process, you're ready to start using the program. Chapter 4 contains instructions for sending, and Chapter 5 has the procedure for receiving.

CHAPTER

4

Sending Fax Messages with WinFax Lite

THERE are two ways to send a fax from WinFax Lite: you can "print" a file to your fax modem from the Windows application where you created the file, or you can use the WinFax Administrator program to send a brief note or a previously created file. In either case, you can add a cover sheet and include additional files, including files created with a different program. For example, you could send a Word-Perfect text file along with a CorelDRAW graphic file in a single fax message. You can send a fax message to a single recipient, or you can "broadcast" it to a list of destinations.

WinFax can send a fax message immediately, or it can delay transmission until a specified time; for example, until the recipients will be in their offices or until the long-distance telephone rates are lower.

This chapter contains complete instructions for sending fax messages from applications and from the WinFax Administrator program. It also includes information about making and using cover pages and scheduling delayed transmission.

Sending a Fax from a Windows Application

The Windows Print Manager treats WinFax as an additional printer connected to your PC. When you're working on a file in your word processor, graphics program, spreadsheet, or other Windows application, you can select WinFax as your printer, and send the file as a fax message.

In order to use WinFax, the Windows Print Manager must be active. Select the Control Panel icon in the Main group, and then double-click on the Printers icon. When the Printers dialog box appears, select the Use

Print Manager check box. If you expect to use WinFax more often than you print copies of files, you can make WinFax your default printer by highlighting WINFAX on COM1 (or whatever serial port is connected to your modem) and clicking on Set As Default Printer. Click on the Close button to close the dialog box.

Follow these steps to send a fax from a Windows application:

1. Use the application's Select Printer or Printer Setup command to make WinFax your current printer. If WinFax is the default printer, you can skip this step.

2. After you've specified WinFax as the current printer, enter the Print command for the current application. Select any options the application's print command requests, and click on OK.

N O T E

Different Windows applications have slightly different Print commands, but in most cases, the command appears in the File menu. If you don't know how to print from a particular application, look in the manual for that application, or press the F1 key to see a Help screen.

3. WinFax will "print" the document using formatting instructions from the application. After you send the document to WinFax, you won't be able to edit the text, use a different font, or make any other changes to the document.

WinFax prints your document by sending it to your modem as a fax message. Use the Fax Send dialog box to specify the name, fax number, and other information.

Working with the Fax Send Dialog Box

When you enter a print command, the application sends the command to the WinFax driver. At this point, you will see the Fax Send dialog box shown in Figure 4.1.

The Fax Send dialog box contains the fields listed below.

Recipient

If you want to send this file to somebody whose name is in your phonebook, click on the Select button to display the Phonebook Details dialog box. (Some versions of WinFax Lite call this the Phonebook Entries dialog box, but the only difference is in the name.) Highlight a name from the list in the Recipients box, or choose a different phonebook to see more names. To see the fax number and some information about the highlighted name, click on the Expand>> button. (Chapter 6 contains information about editing your phonebooks.) When the name of the recipient is highlighted, click on OK. Figure 4.2 shows the expanded Phonebook Details dialog box.

FIGURE 4.1

The Fax Send dialog box

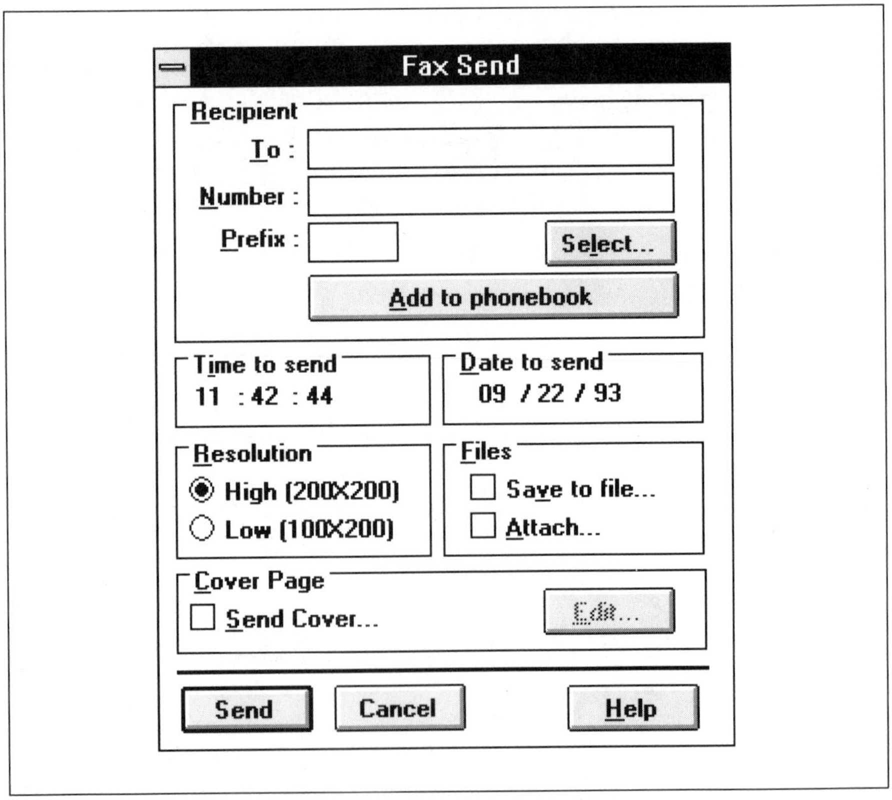

FIGURE 4.2

The expanded
Phonebook Details
dialog box

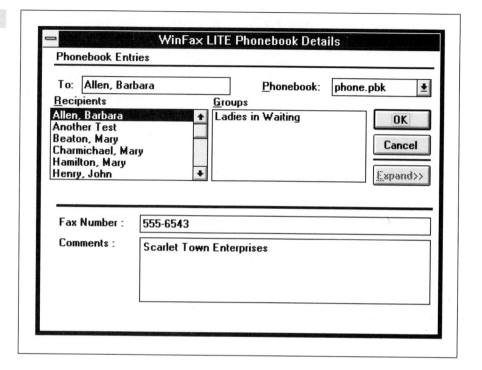

If you want to send a fax to somebody whose name is not in your phone-book, or if you choose not to use a phonebook, you can type the recipi-ent's name in the To field and the fax telephone number in the Number field. To add this name and number to the current phonebook, click on the Add To Phonebook button.

If the Partial Matches option is enabled in the WinFax Phonebook Record dialog box, WinFax will complete the name and fax number of an entry in the current phonebook when you type the first few letters of the name in the To field.

The Prefix field already contains the default prefix that you specified when you completed the WinFax Setup dialog box. If you're using a lap-top or other portable computer away from your home or office, you may need to change or delete the default prefix. You could also use the prefix field to dial an account code or credit card number. To add a pause be-tween the prefix and the fax number, or between digits in the prefix, use a comma (,). The maximum length of a dial prefix is 24 characters.

Time to Send/Date to Send

The default time and date to send are the current time and date. To send a fax as soon as possible, don't make any changes. If you want to schedule delayed transmission, highlight the number of the hour, minute, second, month, day, or year that you want to change. Then use your ↑ or ↓ keys or click on the up or down arrows in the dialog box to change the number. You can move among the hours, minutes, and seconds fields or the month, day, and year fields with the ← and → arrow keys.

Resolution

WinFax uses the Default Resolution unless you choose a different resolution in this field. You can change the default from the WinFax Setup dialog box. In most cases, you should select High Resolution for the best possible image quality. But if you want to hold down your long distance telephone charges, you can choose Low Resolution to reduce the amount of time needed to transmit each page.

Files

Select Save To File... to save this document as a fax attachment file, which you can add to another fax document later.

Select Attach... to add a previously saved fax attachment file to the document you're sending in this fax message.

Cover Page

Click on Send Cover... to send your standard WinFax cover sheet as the first page of this transmission. To make changes to the cover sheet, click on the Edit... button. For information about creating a cover page, read the "Creating a Cover Sheet" section later in this chapter.

After you have completed the Fax Send dialog box, click on the Send button to send the fax. If you want to save the current file as a fax attachment without sending it as a fax right now, select Save To File... and click on the Save button.

When you click on the Send button, the application prints the file to the WinFax Administrator program. At the time and date you specified in the Fax Send dialog box, the administrator places the call and sends the fax message.

When the WinFax Administrator sends a fax message, it displays the progress of the call directly under the menu. If Display Call Progress is active, the same information appears in a dialog box, even if the WinFax Administrator is minimized.

Using Windows Dynamic Data Exchange to Send Faxes Automatically

If your Windows application supports Dynamic Data Exchange (DDE), you can use DDE to exchange commands and data with WinFax. Appendix A contains detailed WinFax DDE specifications.

Sending a Fax Directly from WinFax Lite

Sometimes you want to send a fax that only contains a few sentences or a standard file, such as a catalog sheet or schedule. The easiest way to do this with WinFax is to send directly from the WinFax Lite Administrator.

Starting at the WinFax Lite Administrator screen, select the Fax... option from the Send menu, or click on the Send button in the toolbar. You will see a Fax Send dialog box similar to the one that appears when you send a fax from a Windows application, as shown in Figure 4.3.

If you're not familiar with this dialog box already, read the instructions for using it in the previous section of this chapter.

To send a brief note, select Send Cover and click on the Edit... button. When the Cover Page dialog box shown in Figure 4.4 appears, type your message in the Cover Page Text box.

To send a file that you previously saved as a fax attachment file, select Attach... in the Files box and move the name of the file from the Available Files list to the Transmission List in the Transmit Files dialog box shown in Figure 4.5.

FIGURE 4.3

The Fax Send dialog box

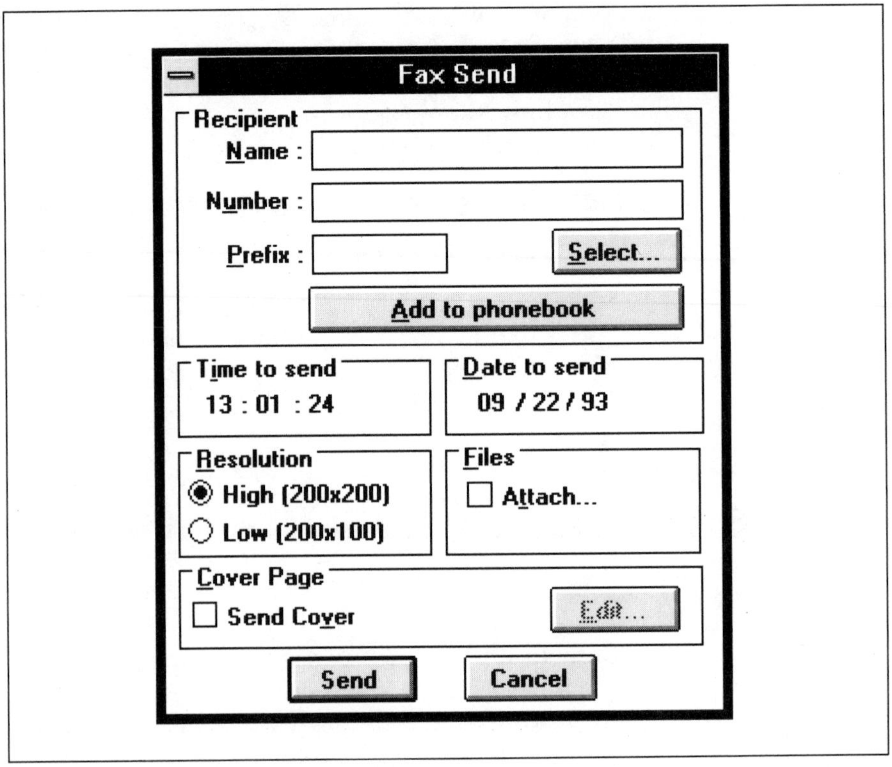

FIGURE 4.4

The Cover Page dialog box

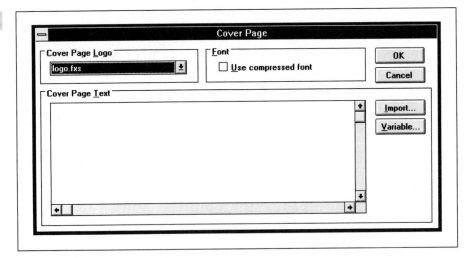

FIGURE 4.5

The Transmit Files
dialog box

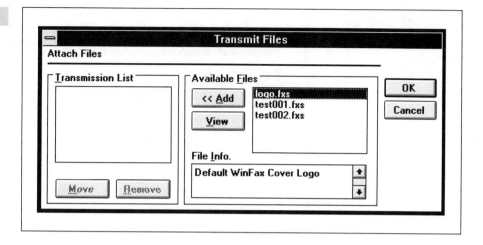

At the time and date specified in the Fax Send dialog box, WinFax will send the cover sheet or file (or both) as a fax. When the WinFax Administrator sends a fax message, it displays the progress of the call directly under the menu.

Combining Files from Different Applications in the Same Fax

If you're using several Windows applications, you may want to combine several files into a single fax message. For example, you might want to include a CorelDRAW or Adobe Photoshop picture with a letter written in Microsoft Word. In other situations, you might want to create a group of standard pages that you will include with custom letters to different recipients.

WinFax uses *attachment files* to store files from applications that you want to save for later faxing. To create an attachment file, print the file from the application, with WinFax specified as the current printer. When the Fax Send dialog box appears, select Save To File... in the Files box. The Save To File dialog box shown in Figure 4.6 will appear.

WinFax will save the document as a file with a .FXS extension and a three-digit page number. Therefore, you're limited to a five-letter name.

FIGURE 4.6

The Save To File
dialog box

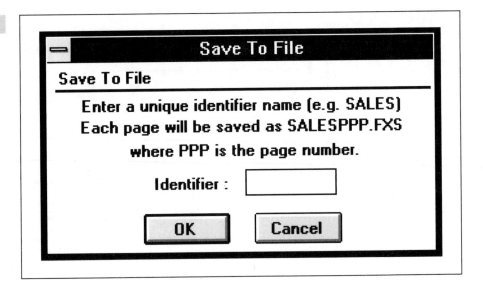

Assign a file name to the .FXS file and click on the OK button to return
to the Fax Send dialog box. Notice that everything except Save To File...
is grayed out. Click on the Save button to save the file.

When you want to send one or more of these fax attachment files with a
document from an application, print the document from the application.
When the Fax Send dialog box appears, click on Attach... in the Files box.
WinFax will display the Attach Files dialog shown in Figure 4.7.

FIGURE 4.7

The Attach Files dialog
box

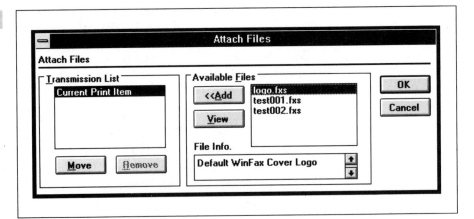

ll me hen you come to a decision.

When you highlight an attachment file, the dialog box displays a brief description of that file in the File Info. box. If you need more information, click on the View button to display the highlighted file in the WinFax Image Viewer.

The Transmission List shows the items that WinFax will include in this fax message, in the order they will be sent. The Current Print Item is always the first item in the list. To add a file to the Transmission List, highlight the filename and click on the <<Add button. Repeat the process for each additional page you want to send in this message.

To remove a file from the Transmission List, highlight that listing and click on the Remove button. It's not possible to remove the Current Print Item from the list.

To change the order in which WinFax will send attached files, highlight the name of the file you want to move and click on the Move button. When you move your mouse into the Transmission List box, it becomes an insertion pointer instead of the usual arrow. Place the pointer where you want the highlighted file to move, and click on the left mouse button.

When the Transmission List shows all of the files you want to include in this fax in the correct order, click on OK.

Creating a Cover Sheet

It's common practice to send a cover sheet as the first page of a fax message. The cover sheet usually identifies the person to whom the fax is directed, the name and fax number of the originator, the date, and the number of pages in the transmission. When a business receives its faxes on a shared fax machine, the name on the cover sheet makes it easy to know who should get the message. By checking the total number of pages on the cover sheet, the recipient can make sure the message is complete.

When you use a WinFax cover sheet, you can automatically insert the names of the originator and recipient, the date and time of transmission, and the number of pages. In addition, you can include a company logo or other graphic image and up to 45 lines of text, or up to 79 lines without a logo or other graphic. Figure 4.8 shows a typical cover sheet.

FIGURE 4.8

A typical coversheet created with WinFax

FACSIMILE COVER PAGE		
To:		From:
Time: 13:04:24		Date: 11/23/93
Pages (including cover): 1		

This is a sample WinFax Lite Cover Page.

To create a WinFax Cover sheet, select Define Cover Page from the Win-Fax Administrator's Send menu or File menu. The Cover Page Setup dialog box shown in Figure 4.9 will appear.

The standard WinFax cover sheet has three sections: the information block, the logo, and the text. You can include or omit any of these sections.

The Cover Page Information Block Template

The information block contains the names of the recipient and sender, the date and time of transmission, and the total number of pages. WinFax includes this information block as part of the cover sheet when Enable Cover Page Information Block is active. If you don't want your cover sheet to include an information block, click on the box next to Enable… to remove the X from the box.

The field headings in the information block are pretty standard, but you can change them if you want to. For example, you could personalize your cover sheet by changing the title from "FACSIMILE COVER PAGE" to "ANOTHER FAX FROM SUE'S SHOES." If you're sending a fax to non-English speakers, you can also use this feature to translate the cover sheet headings to another language. To change a field heading, move to the box that contains it and edit the text. You can return to the standard headings by clicking on the Default button.

The Cover Page Logo File

In addition to the information block, a WinFax cover sheet can also include your company's logo or some other graphic image. The original graphic may be stored in any graphics file format that you can read from a Windows application. To use a graphic file as a logo, convert it to a .FXS file by printing it to the WinFax printer driver from the original application. In the Fax Send dialog box, select Save To File and assign a five-letter file name. When you click on the Save button in the Fax Send dialog box, WinFax will create a .FXS file with the name you specified. For a more detailed explanation of this procedure, read the "Combining Files from Different Applications in the Same Fax" section earlier in this chapter.

To place the logo on your cover sheet, open the drop down menu in the Cover Page Logo File box and select the name of the file that contains the logo.

The Default Cover Page Text

The cover page text may be either standard text imported from an ASCII text file, or a specific message for this cover sheet. You can increase the maximum amount of text from 29 lines to 45 lines (or from 51 lines to 79 lines if you don't include a logo) by selecting Use Compressed Font.

The cover page text can include variable items that automatically change in each message. Since variables use a dollar sign ($), you must use two dollar signs ($$) when you want to include a dollar sign in your text. For example, to include this sentence:

> Total cost for this contract is $1,450.

type it like this:

> Total cost for this contract is $$1,450.

To use the same text on every cover sheet, click on Use Default Cover Page Text File to place an X in the box, and then click on the Select button. Choose the text file you want to use, and click on OK. The file name is now in the box under Use Default Cover Page Text File.

To use a different text in every cover sheet, make sure there is no X in the box next to Use Default Cover Page Text File. Even if you specify a default text file, you can edit or remove the text later for individual faxes.

Making a Custom Cover Sheet

WinFax automatically adds a cover sheet to every fax message if the Send Cover option is active in the Fax Send dialog box. Unless you instruct the program to change it, WinFax will use your standard cover sheet. To change the cover sheet on an individual transmission, click on the Edit button next to Send Cover to display the Cover Page dialog box shown in Figure 4.10.

FIGURE 4.10

The Cover Page dialog box

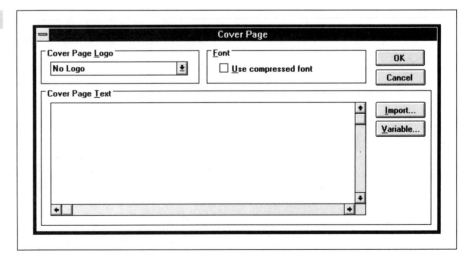

The Cover Page dialog box will appear, with the file name of the standard logo and the default text in their respective boxes. Click on the down arrow in the Cover Page Logo box to select a different logo.

If there's a default text, it appears in the Cover Page Text box in reverse type. To delete the entire text block, press the Backspace key once. To add text to the default text block, use the mouse to place the cursor at the location where you want to edit the text. To delete text, drag the cursor to highlight the text you want to remove, and press the Delete key.

To use the contents of an ASCII text file in your text block, click on the Import button and choose the file from the file selector. WinFax will insert the text file at the cursor.

A variable in a text block automatically prints specific information about the current fax message. If you send the same cover sheet with variables with several fax messages, WinFax will replace the variable with different information in each cover sheet.

To use a variable in your text block, move the cursor to the location where you want to insert the variable, and click on the Variables button. The Insert Variable dialog box in Figure 4.11 will appear. Choose the variable you want to use, and click on OK.

If you schedule delayed transmissions, remember to leave your computer on and the WinFax Administrator active until the scheduled transmission time. If the computer is not running, or if the WinFax Administrator is not active at the time you selected for a delayed transmission, WinFax will send the fax the first time you start the Administrator after the specified time.

FIGURE 4.11

The Insert Variable dialog box

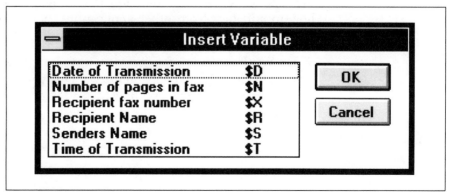

Delaying Transmission of a Fax

There are lots of reasons to prepare a fax message now for transmission later. For example, you can save money by sending long-distance faxes at night, when long-distance rates are lowest. Or if you're sending a fax to someone who's out of the office, you might want to delay transmission until that person returns.

WinFax makes delayed transmission easy. The Fax Send dialog box includes Time To Send and Date To Send fields that default to the current time and date. To schedule later transmission, change the time or date field. The scheduled fax will remain in the WinFax Administrator's event list as a "Pend." item until WinFax actually sends it.

To change a scheduled time or date, highlight the call in the event list and click on the Resched button or select Send ▶ Reschedule. The Reschedule Event dialog box includes Time and Date fields where you can make changes.

To cancel a scheduled transmission, highlight the event in the event list and click on the Remove button.

If you try to shut down the WinFax Administrator while it is holding a fax message for future transmission, it may display a warning message.

You can send faxes directly from Windows applications by selecting the application's Print command and choosing WinFax as your printer. In this chapter, we've explained how to complete the Fax Send dialog box, how to add files from other applications, and how to create a custom cover sheet for your faxes.

When WinFax sends a fax message, it stores information about the message in the Send Log. In Chapter 6, we'll tell you how to use the Send Log to print or display a fax, and how to send another copy of a fax, either to the original recipient or to a different destination.

CHAPTER

5

Receiving
Fax Messages in
WinFax Lite

YOU CAN use WinFax to receive faxes in the background when Windows is active, or as a memory-resident DOS program when you're not using Windows. This chapter explains how to set up your computer to receive fax messages and describes the messages you will see on your screen when an incoming fax call arrives. We'll explain how to display and print faxes in the next chapter.

To receive faxes with WinFax, your computer and modem must both be turned on. This seems obvious when you think about it, but most of us turn off the computer at the end of the day as a matter of habit. If you've given your fax number to other people, they'll expect to be able to send you a fax at any time of the day or night, so you must remember to leave both devices on at all times. Since WinFax runs in the background, you can turn off the monitor to reduce power consumption and prevent burning an image on your screen, but the system unit and modem must remain on.

Receiving Faxes when Windows Is Running

WinFax can use the WinFax Administrator program to receive faxes whenever Windows is active. You can configure WinFax to either automatically answer the telephone and start receiving, or let the telephone ring until you answer manually. When WinFax receives a fax message, it can display the received message, automatically print it, display a notice that a new message has arrived, or simply store the message and wait for you to run the WinFax Administrator program. The Receive Setup dialog box specifies the currently active receive options.

Configuring Receive Setup

To change the WinFax receive configuration, start the WinFax Administrator program, and select Receive Setup from the Receive menu. The Receive Setup dialog box shown in Figure 5.1 will appear.

FIGURE 5.1

The Receive Setup
dialog box

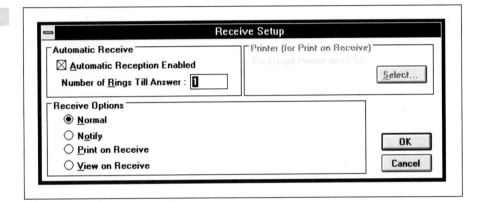

The Receive Setup dialog box has the following three sections:

Automatic Receive

Select Automatic Reception Enabled to instruct WinFax to automatically answer the telephone when it rings.

When automatic reception is not enabled, you can use WinFax to receive a Fax message by choosing Manual Receive from the WinFax Administrator Receive menu.

There are two ways to use automatic reception. If you have a dedicated fax line separate from the telephone number you use for voice calls, select Automatic Reception Enabled and set the number of rings to 1. Assuming your computer is turned on and Windows is active, WinFax will answer every call as soon as the telephone rings.

On the other hand, if you share the same telephone line for voice calls and fax calls, you may want to set the number of rings to a number large enough to permit you to pick up the telephone handset (allow about ten rings per minute) before the computer answers it for you. If you're not there to answer the call, WinFax will answer after the specified number of rings. If you happen to pick up a fax call and hear fax tones, you can use Manual Receive to accept the fax message.

Automatic reception doesn't work when Windows is running DOS or a DOS application as a full screen display, but it will accept calls if DOS is running within a window on the Windows Desktop. If a fax call comes in when a full-screen DOS display is active, either press Alt-↵ to convert the full screen to a window, or use the memory-resident WinFax DOS program.

Printer

Use the Select... button to change the printer that WinFax will use to print incoming fax messages when the Print On Receive option is active.

Receive Options

WinFax offers four possible responses when it receives a call:

Normal   When this option is active, WinFax receives calls in the background while you are using other Windows applications or the Windows desktop. Under the Normal option, WinFax stores received fax messages on disk as fax "events." You must look at the WinFax Administrator window to find out if any new fax messages have arrived at your computer.

This may sound like a minor nuisance, but it's really pretty simple compared to walking across the office to a stand-alone fax machine. If you don't think you'll remember to check the WinFax Administrator, choose the Notify option instead.

Notify   When this option is active, WinFax displays the pop-up *WinFax LITE Fax Received* message in Figure 5.2 whenever it receives a new fax.

See you after vacation.

FIGURE 5.2

The *WinFax LITE Fax Received* message box

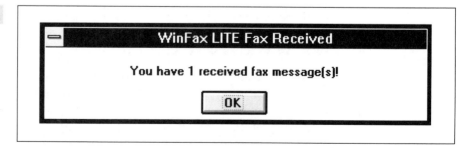

The *Fax Received* message shows the number of messages received since the last time you clicked on OK. Clicking on OK returns the message counter to zero.

When this message pops up, you have four options:

- You can move it out of your way but leave it on the screen as a reminder that you have a new fax message.
- You can start the WinFax Administrator to display or print the new fax message.
- You can use the Control-menu box (in the upper-left corner of the pop-up message) to close the message window without resetting the message counter.
- You can click on OK to reset the counter and close the message window.

Print On Receive   When this option is active, WinFax sends all fax messages to your printer as soon as it receives them. Use the Select... button in the Printer box (for Print On Receive) to specify the printer.

Before you select this option, remember that WinFax takes a l-o-n-g time to print each page. Unless you are sure you're going to need a hard copy of every incoming fax, View On Receive might be a better choice. On the other hand, if you've got a spare printer and an extra printer port on your computer, you could set up WinFax to automatically turn that printer into a "receive only" fax machine, without interfering with the other work you do on the same computer.

It's possible to specify WinFax as your printer, but it's hard to think of a reason you'd want to do that.

View On Receive   When this option is active, WinFax automatically starts the WinFax Image Viewer whenever it receives an incoming fax message. Chapter 6, "Viewing and Printing Fax Messages in WinFax Lite" explains how to use the WinFax Image Viewer.

View On Receive waits until WinFax has received the entire message before it starts the Image Viewer, so you can't watch each page as it comes into your computer.

When automatic reception is active, an icon that resembles a fax machine is visible under the menu bar of the WinFax Administrator window. You can call the Receive Setup dialog box by double-clicking on the icon. When WinFax sends start-up codes to the modem, an hourglass appears inside the icon. When the hourglass disappears and a lightning bolt appears, WinFax is ready to receive.

Receiving Calls

In order to receive faxes, either the Windows WinFax Administrator or the DOS memory-resident WinFax program must be active. This section explains the WinFax Administrator; we'll explain the DOS program later in this chapter.

There are two sets of WinFax options that affect the way the program receives calls: Automatic Reception Enabled in the Receive Setup dialog box, and Display Call Progress in the WinFax Setup dialog box.

When you have a separate telephone number for your fax line, or if you have a line sharing device that sends fax calls to your modem and voice calls to a telephone, you'll want to configure WinFax for automatic reception. If automatic reception is active, WinFax will automatically answer incoming calls to the telephone line connected to your modem after the number of rings you specified in the Setup dialog box.

N O T E

It's important to remember that WinFax only answers calls and receives faxes in Windows when the WinFax Administrator program is running. Therefore, it's a good idea to set up Windows to start the WinFax Administrator automatically when you start Windows, by moving the WinFax icon to the Startup group (you can move an icon from one group to another by dragging it with your mouse). If you don't want the WinFax Administrator screen to appear every time you start Windows, select the WinFax icon, open the Properties dialog box in the Program Manager File menu, and click on the Run Minimized check box.

If you use the same telephone line for voice and fax calls without a line sharing device, don't activate automatic reception. When the telephone rings, answer it as you would any other telephone call. If you hear a high-pitched tone, the call is a fax message. To receive it, double-click on the WinFax icon, and then select Manual Receive from the Receive menu.

When Display Call Progress is enabled, the pop-up Fax Reception Status dialog box shown in Figure 5.3 appears while WinFax receives incoming fax messages. In this dialog box, the Operation line shows the current transmission stage of the call. Calling Station shows the station identifier of the calling fax machine, and Page shows the current page that WinFax

FIGURE 5.3

The Fax Reception Status dialog box

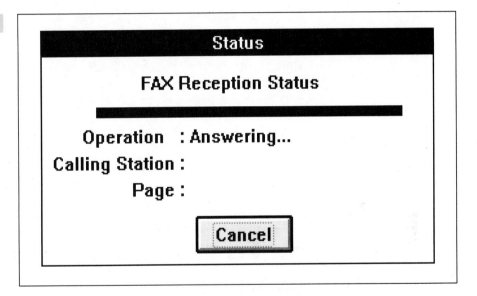

is receiving. To break off the connection before the transmission is complete, click on the Cancel button.

When Display Call Progress is not enabled, fax status information appears in the WinFax Administrator work area, as shown in Figure 5.4. As in the Fax Reception Status dialog box, you can stop an incoming fax transmission by clicking on the Cancel button.

When the transmission is complete, WinFax stores the received fax message and adds it to the top of the event list in the WinFax Administrator

FIGURE 5.4

Fax status information in the WinFax Administrator work area

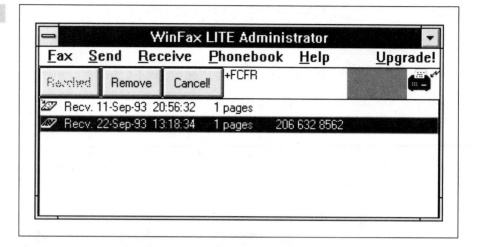

window. If you specified the Print On Receive option in the Receive Setup dialog box, WinFax will send the received fax message to your printer as soon as the transmission is complete; if you specified View On Receive, it will automatically start the Image Viewer and display the first page of the newly received fax message.

Chapter 6, "Viewing and Printing Fax Messages in WinFax Lite" explains how to print or display a received message.

Receiving Faxes when Windows Is Not Running

Since the WinFax Administrator is a Windows application, it can't receive incoming fax messages when Windows is not running. Therefore, Delrina has included a memory-resident DOS receive program in the WinFax Lite package. If you always run Windows, you don't need this program, but if you expect to receive fax messages when you're not using Windows, it may be worthwhile to add it to your AUTOEXEC.BAT file. The memory-resident DOS WinFax program uses about 32K of memory.

The DOS WinFax Command

From the DOS prompt, you can display the DOS WinFax syntax by typing the following (you must be in the WinFax directory or include the path):

WINFAX

This is the command that starts the WinFax TSR:

WINFAX [options] -P*comport* -W*path*

-P*comport*	The COM port where WinFax will receive fax messages. Use the number of the serial port in place of *comport*. For example, if your modem is connected to COM2, use -P2.
-W*path*	The WinFax directory's path. For example, if the WinFax directory is C:\WINFAX, use -WC:\WINFAX.

Options:

-S*x*	Speaker volume. Use -S0 for low volume, -S1 for medium volume, or -S2 for high volume. The default is -S1.
-C*x*	Speaker control. Use -C0 for "speaker always off," -C1 for "speaker on until connected," or -C2 for "speaker always on." The default is -C1.
-R*x*	Number of rings until the modem answers. The default is 2.
-Q*x*	The hardware interrupt for the serial port connected to your modem. Normally, DOS uses IRQ 4 for COM 1 and COM 3, and IRQ 3 for COM 2 and COM 4, so you should use -Q4 if your modem is connected to COM 1 or COM 3, or -Q3 if your modem is connected to COM 2 or COM 4.

-B*x* The base address of the serial port connected to your modem. *x* may be 3F8 (COM 1), 2F8 (COM 2), 3E8 (COM 3), or 2E8 (COM 4).

-M*string* The modem initialization string. Look in your manual for the specific string your modem requires, or copy the Init. String from the WinFax Fax/Modem Setup dialog box.

-N Display the progress of incoming calls as the computer receives them. When the -N switch is active, WinFax places a status message in the upper-right corner of the screen as it receives an incoming fax message. Figure 5.5 shows the status message on a DOS screen. When the -N switch is not active, WinFax places a smaller, flashing "smiley face" indicator in the upper right corner after it receives a fax message.

FIGURE 5.5

A WinFax status message on a DOS screen

```
                                                    Fax Reception
C:\>
C:\>cd winfax

C:\WINFAX>winfax -u
WINFAX TSR is not loaded!

C:\WINFAX>winfax -p4 -wc:\winfax -n
WINFAX TSR (c) Delrina Technology Inc. 1993
Press Left-Shift and Enter for manual reception

C:\WINFAX>
```

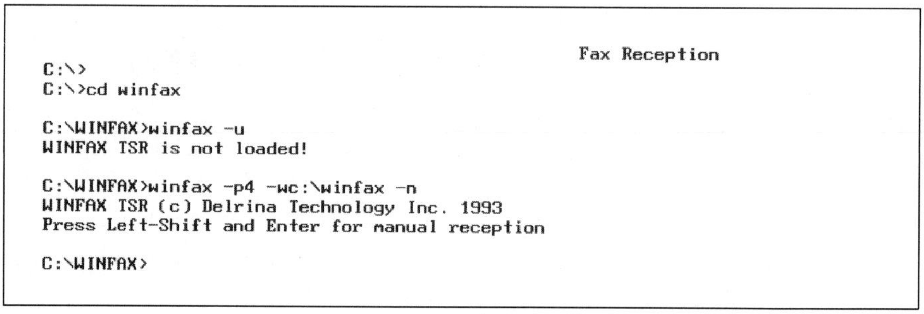

-I*ident* The station identifier that WinFax will return to the distant fax machine to identify your line. Many fax machines and programs (including WinFax) use the Station Identifier in fax activity lists or displays. Most people use their telephone number as their Station Identifier.

When you enter the DOS WinFax command, or if it is part of your AUTOEXEC.BAT file, the program starts in automatic reception mode. In automatic mode, WinFax automatically answers incoming calls after the number of rings specified in the command line.

To switch to manual reception mode, hold down the left Shift key and press ↵. When the modem receives an incoming call, press the same pair of keys to answer and start receiving a fax message. In order to use Win-Fax in manual mode, you must have a telephone instrument plugged into your modem, or installed as another extension on the same line, since WinFax does not tell you that an incoming call is ringing your line.

When the transmission is complete, WinFax stores the received fax message in the WINFAX\DATA subdirectory and adds it to the top of the event list in the WinFax Administrator. You must start Windows and use the Windows WinFax program to display or print received fax messages.

Removing WinFax from Memory

To remove the WinFax DOS TSR from your system's memory, use this command:

    WINFAX -U

Checklist

In order to automatically receive fax messages with WinFax, these conditions must be true:

✓   Modem on

✓   Computer on

✓   Windows running

✓   WinFax Administrator running

✓   Automatic Reception Enabled (Receive Setup dialog box)

Or:

✓ Modem on

✓ Computer on

✓ Windows *not* running

✓ DOS memory-resident WinFax program active in automatic reception mode

In this chapter, we set up WinFax and received a fax message. The next step is to examine the message, either by displaying it on your screen or printing a copy. The next chapter contains detailed information about both of these procedures.

CHAPTER

6

Viewing and Printing Fax Messages in WinFax Lite

EVERY TIME WinFax sends or receives a fax message, it stores that message as a *fax event* in the WINFAX\DATA subdirectory. Fax event files have the file extensions .FXS, .FXD, or .FXR. You can display a fax event on your screen, print a paper copy, convert it to a graphic file format for use with another application, or transmit it through your modem as a fax message. This chapter explains how to use the WinFax Image Viewer, how to export fax events to graphic formats, and how to print copies of fax messages. You can find instructions for sending fax events as fax messages in Chapter 4.

Displaying Fax Messages on Your Screen

To display a fax event on your screen, use the WinFax Image Viewer. There are several ways to start the Image Viewer:

- Select View in the WinFax Lite Administrator Fax menu.
- Click on the View... button in any WinFax dialog box where it appears.
- Specify View On Receive in the Receive Setup dialog.

When the Image Viewer starts from a dialog, it displays the file or fax event that had been highlighted when it was started. When the Image Viewer starts because WinFax has received a fax message, it displays the first page of that message.

As you can see in Figure 6.1, the Image Viewer includes five menus and a button bar.

FIGURE 6.1

The WinFax Image
Viewer screen

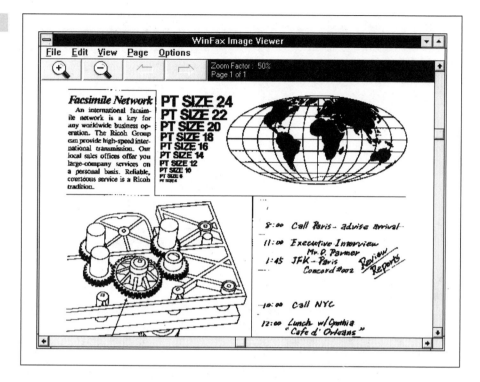

In most cases, the Image Viewer starts up with the fax event that you want to see already loaded. To display a different fax event, select File Open... and select a file from the Open Image File dialog box shown in Figure 6.2. The default file extension is .FXS, but you can also select .FXR files and .FXD files. In general, however, it's easier to request transmitted faxes (.FXD) with the Send Log and received faxes (.FXR) with the Receive Log.

Changing the Size of the Image on Your Screen

The Image Viewer can display an entire page, a magnified portion of the page, or up to eight pages of a multipage fax message at the same time. Use the View menu to specify the size of the image on your screen. As Figure 6.3 shows, you can select "zoom factors" of 25%, 50%, or 100%, or fit the entire page into the viewing area by selecting Full Page. Figures 6.4 and 6.5 show the same file at different zoom factors. The current zoom factor value appears inside the button bar.

FIGURE 6.2

The Open Image File
dialog box

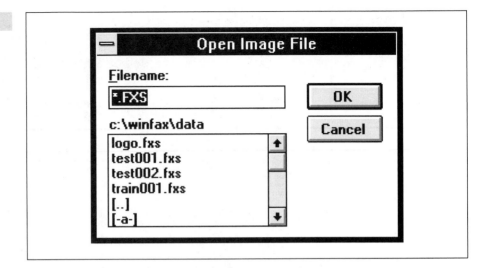

FIGURE 6.3

The Image Viewer
View menu

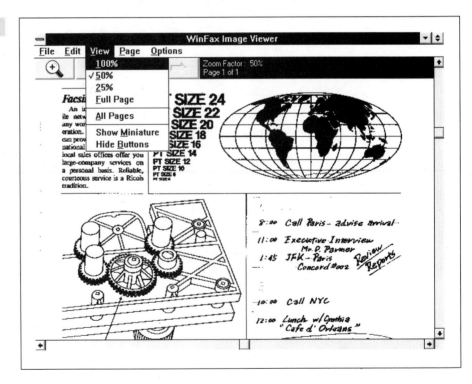

FIGURE 6.4

Image Viewer with a 50% zoom factor

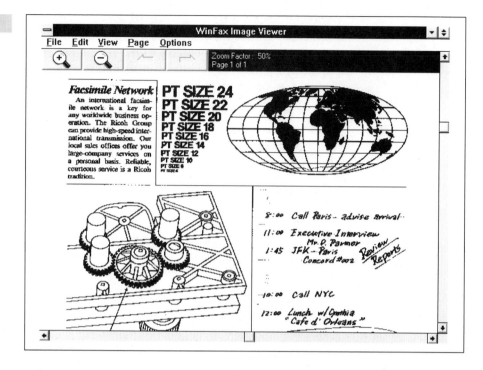

FIGURE 6.5

The same image viewed with 100% zoom factor

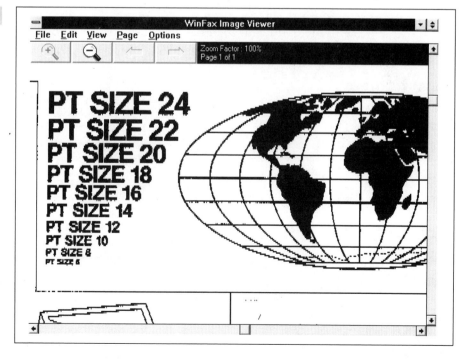

You can also use the two magnifying glass buttons to increase or reduce the view. Choose the + button to change the image to the next larger zoom factor, or the − button to the next smaller zoom factor. If only one magnifying glass icon button is visible, the image is already at the largest or smallest zoom factor.

When you click on the + button, the Viewer displays a box in the display area that shows the size of the next larger image. Use your mouse to drag the box to the section of the image that you want to see, and click on the left mouse button to see the enlarged image.

You can also superimpose a small picture of the entire current page over the larger image by double-clicking when your mouse is inside the display area, or selecting Show Miniature in the View menu. Figure 6.6 shows a miniature view. There's a box in the miniature view that shows the location of the main image within the current page. By dragging the box around the miniature view, you can move to a different location on the page.

To make the miniature view disappear, either double-click on the bar in the upper-left corner, or select Hide Miniature from the View menu.

Moving around in a Document

When you receive a fax message with more than one page, you can use the options in the Page menu or the arrow buttons to display a different

FIGURE 6.6

A miniature view of a fax message

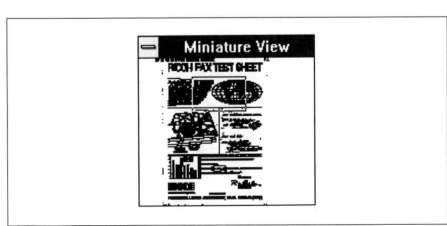

page. The current page number and the total number of pages appear inside the button bar. When you choose the Go To Page... option, the program displays a dialog box that lets you specify the number of the page you want to see. Figure 6.7 shows the Go To Page dialog box.

Rotating the Image

If an image appears in the Image Viewer upside down or sideways, you can use the items in the Options menu to rotate the image to make it easier to read. Figure 6.8 shows the Options menu.

Select Rotate Current Page to display a submenu with four choices; choose 90 Degrees or 270 Degrees to rotate an image that was received sideways; or 180 Degrees to fix an upside down image. Choose Original Rotation to cancel a previous rotation command. Figure 6.9 shows an image rotated 90 degrees.

If all of the pages in a fax message are upside down, you can use the Flip All Pages option to fix the whole message at one time.

FIGURE 6.7

The Go To Page dialog box

```
┌─────────────────────────────────────┐
│   ┌───────────────────────────┐      │
│   │ ▬    Go To Page           │      │
│   ├───────────────────────────┤      │
│   │      Total Pages : 2      │      │
│   │                           │      │
│   │   Go to page :  ┌───────┐ │      │
│   │                 │ 2     │ │      │
│   │                 └───────┘ │      │
│   │  ┌────────┐   ┌──────────┐│      │
│   │  │  OK    │   │  Cancel  ││      │
│   │  └────────┘   └──────────┘│      │
│   └───────────────────────────┘      │
└─────────────────────────────────────┘
```

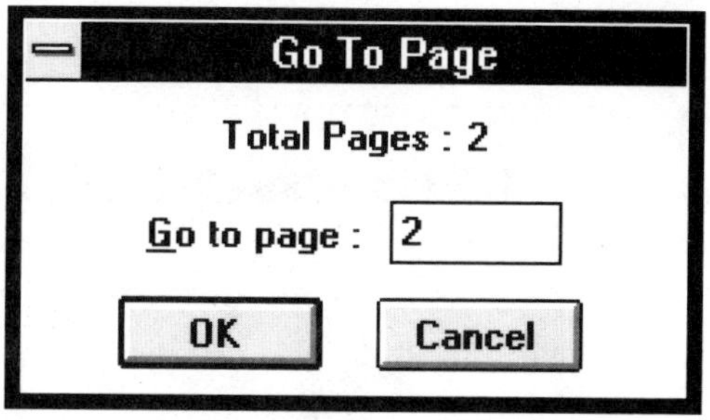

Printing Fax Messages

In order to print a copy of a fax message or a fax attachment, you must first use the Image Viewer to display it on your screen. To print the document currently displayed in the Viewer, choose Print... from the File menu. The Print Page(s) dialog box shown in Figure 6.10 will appear.

FIGURE 6.8

The Options menu

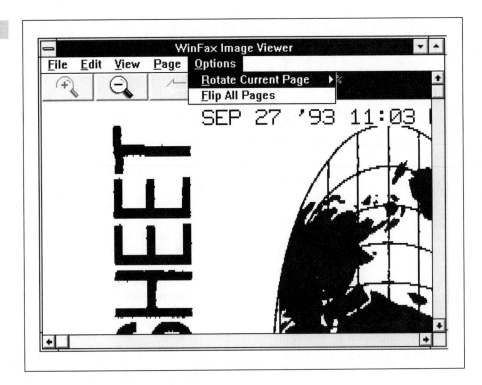

FIGURE 6.9

An image rotated 90 degrees

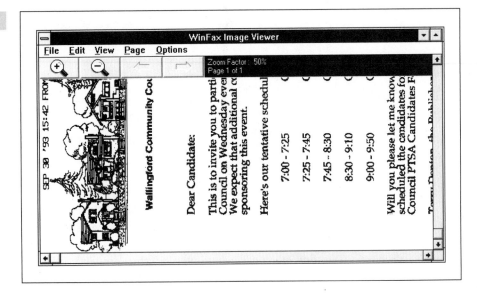

FIGURE 6.10

The Print Page(s)
dialog box

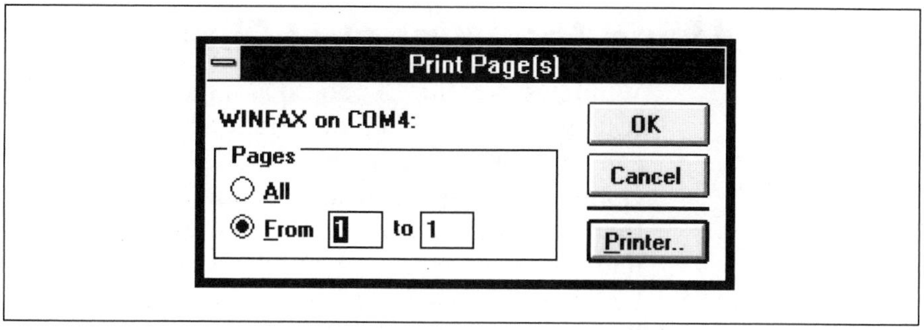

The Print Page(s) dialog box shows the currently selected printer. To use a different printer, click on the Printer... button and select the printer you want to use. To change the default printer without printing anything right now, use the Select Printer... option in the File menu.

To print the entire document, select All in the Pages box. To print specific pages, select From and enter the first and last page number you want to print.

It's possible to select WinFax as your printer and use this technique to send the document in the Viewer as a fax. In most cases, it's a lot easier to use the methods explained in Chapter 4 instead.

Moving Fax Images to Other Applications

One of the strengths of the Windows environment is the ability to move information between different applications. WinFax treats fax messages as graphics files, so it's not possible to send text directly to a word processor, but you can use the Windows Clipboard to copy an image into another Windows program. You can also use the Export command to convert one or more pages of a fax message to a standard graphics format that another application will recognize, and to use the tools in a graphics program to improve the appearance of a received fax image.

Using the Windows Clipboard

Windows uses the Clipboard as a temporary storage area for transferring text, graphics, or other information between applications. In WinFax, you can copy all or part of a page to the clipboard and place it in a word processor or desktop publishing document, or any other Windows program that uses graphic images. For example, if your branch office faxes you a map for inclusion in a press release, you could use the clipboard to transfer the map to your word processor. The WinFax clipboard commands are in the Image Viewer's Edit menu shown in Figure 6.11.

FIGURE 6.11

The Image Viewer Edit menu

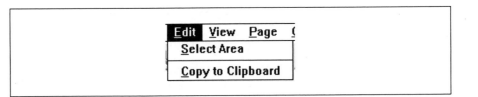

To define the portion of a fax image you want to copy to the clipboard, first set the Image Viewer's zoom factor so that the entire portion to be copied is visible on your screen. Now choose the Select Area option in the Edit menu and move the cursor to a corner of the region you want to copy to the clipboard. As shown in Figure 6.12, hold down the left mouse button and drag your mouse until the box on the screen encloses the image you want to move, and then release the mouse button.

After defining the area to be copied, click on Copy To Clipboard in the Edit menu. Open the application where you want to use the fax image, and use that program's Paste or Import From Clipboard command to place the image in a document.

Converting Fax Events to Graphic Formats

The Windows Clipboard is a useful tool for moving a small image to another application, but WinFax includes another method for transferring entire documents. The Export... command in the Files menu makes it possible to convert one or more pages of a received fax message to standard .PCX and .TIF graphics formats that many other applications can recognize.

Drag the cursor to define an image to copy to the Clipboard.

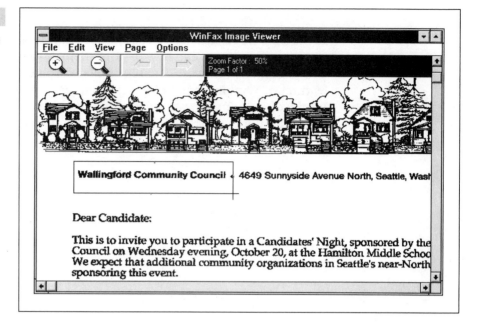

To convert a fax message, display any page of the message in the Image Viewer and select Export... in the File menu. Figure 6.13 shows the Export Fax dialog box. Choose the new file format, enter the name you want to assign to the new file in the File prefix box, and select the directory where you want to store the converted image. If the original fax message is longer than one page, either choose All pages or specify the first and last page you want to convert. Finally, click on OK to start the conversion.

The Export Fax dialog box

In the last three chapters, we've explained how to use WinFax to send and receive fax messages, and how to print them or view them on your screen. The information in these chapters is all you need to understand the basic features of WinFax Lite. In the next chapter, we'll tell you how to use the WinFax Lite Administrator to manage your faxes more easily, and how to work with phonebooks and the WinFax Send and Receive logs.

It's a whale
of a job.

CHAPTER

7

Controlling WinFax Lite with the Administrator Program

IN CHAPTERS 4 and 5, you learned how to use the WinFax Lite Administrator program to send and receive faxes. In this chapter, we'll pull all the Administrator tasks together, explain how to use the Administrator to perform several housekeeping tasks, and give you a general "road map" of the Administrator screen.

Delrina has released at least three different versions of WinFax Lite, each with a slightly different Administrator screen. All three versions have the same set of commands, but the commands are arranged into menus differently. In WinFax Lite 3.0, the Administrator also has more options in the button bar. Don't panic if the Administrator screen you see doesn't look exactly like the pictures in this book. The commands are the same, and they produce the same results.

Since Delrina wants you to upgrade from WinFax Lite to WinFax Pro, the only way to obtain a newer version of Lite is to buy a new modem that has a more recent version of Lite packaged with it. But since all three versions include all the functions you need to send, receive, and manage faxes, there really isn't any good reason to replace the version of WinFax Lite that came with your modem.

If you've moved the WinFax Lite icon to your Windows Startup group, WinFax will automatically load every time you start Windows. At the end of the startup routine, the WinFax Administrator screen should appear on your screen.

If WinFax is not in your startup group, or if the program has been configured to Run Minimized (in the Properties... dialog box under the Files menu), you can display the WinFax Administrator screen by clicking on the WinFax Lite icon.

Figure 7.1 shows the WinFax Administrator screen.

FIGURE 7.1

The WinFax
Administrator screen

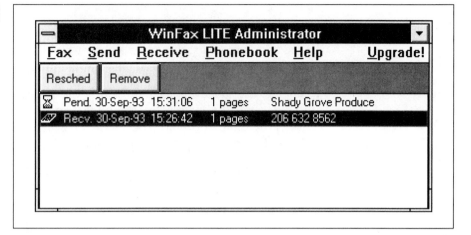

Managing Fax Events

WinFax treats every incoming and outgoing fax message as a "Fax Event." The WinFax Administrator displays a current list of events in the lower portion of the screen. You can use the event list to see information about outgoing fax messages currently in the queue and about messages that you previously scheduled for later transmission. You can remove a fax event from the queue, change the scheduled time of transmission, or cancel the transmission entirely. You can also view and print recently received faxes and create a description of a received fax, which WinFax will store with the fax itself in the Received Fax Log.

The event list identifies fax messages scheduled for transmission as *Pend.* (for *Pending*) with an hourglass icon. It identifies received faxes as *Recv.* with an icon that shows a piece of paper. Each event listing includes the date and time the transmission is scheduled to take place or that the inbound fax was received, the number of pages, and the name or fax number of the originator (for received faxes) or destination (for scheduled outgoing faxes).

For more information about any event on the list, double-click on the event listing, or highlight it and select Event Information... from the Fax menu. A Pending Event Information or Receive Event Information dialog box will appear.

To start the Image Viewer and display the first page of the highlighted fax event, select View... from the Fax menu.

To see a list of fax attachments files (with an .FXS file extension) that are part of the highlighted fax event, select Attachments... from the Fax Menu.

To change the scheduled transmission time or date, click on the Resched. button. When the currently scheduled transmission time and date appears in the dialog box shown in Figure 7.2, select either the time or date and use the arrow keys to make changes.

To cancel a scheduled transmission entirely, highlight the event you want to cancel and click on the Remove button.

FIGURE 7.2

The Reschedule Event dialog box

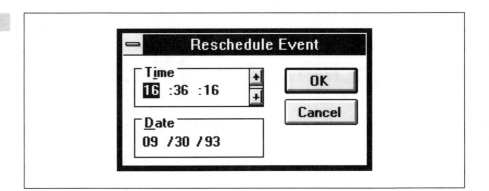

Understanding the Pending Event Information Screen

As Figure 7.3 shows, the Pending Event Information screen contains information about an individual fax message, including the telephone number that WinFax will dial at the scheduled time and other information about the scheduled transmission. You can display the same information about completed transmissions by selecting a listing in the Send Log. Since WinFax uses similar screen for both pending events and completed transmissions, it shows zeroes in the fields that describe the number of pages sent, the duration of the transmission, and the number of retries.

To display a list of fax attachment files, click on the Files button. If there are no files attached, it is not possible to select this button.

The following is intended to drive home point.

FIGURE 7.3

The Pending Event
dialog box

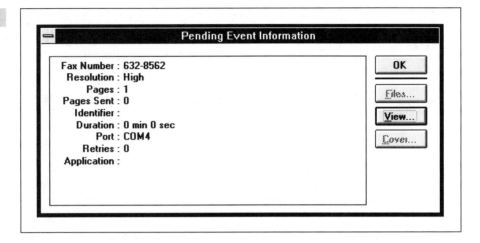

To display the pending fax message, click on the View button.

To display information about the cover sheet attached to this fax message, click on the Cover button.

Understanding the Receive Event Information Screen

Figure 7.4 shows the Receive Event Information dialog box. It identifies the source, resolution, size, and duration of a fax message that WinFax has received. The Identifier field shows the station identifier of the fax machine or computer that originated the message.

FIGURE 7.4

The Receive Event
Information dialog box

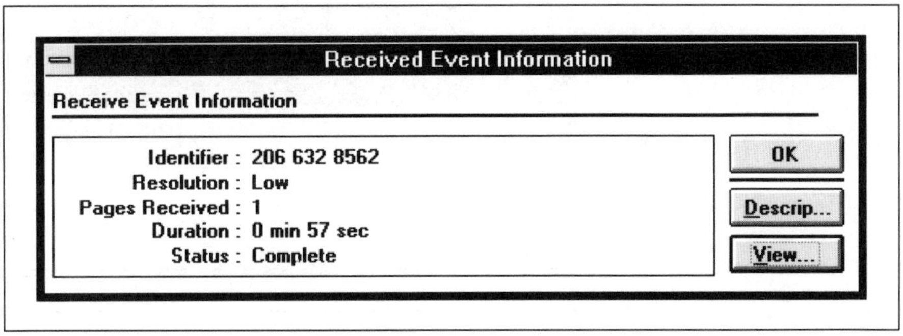

To assign a brief description to this message, click on the Descrip... button. The same description will appear in the receive log.

To display or print this message, click on the View... button.

Observing Fax Message Progress

As WinFax sends and receives faxes, it displays the current process of the call in the WinFax Administrator window, just above the event list. Figure 7.5 shows the WinFax Administrator screen with an active status message.

The same status messages appear in the Fax Transmission and Reception Status dialog box that appear if you made Display Call Progress active in the WinFax Setup dialog box. The status messages appear in the Administrator even if Display Call Progress is not active.

You can stop a fax transmission in progress by clicking on the Cancel! button, which is only visible when WinFax is sending or receiving a message.

Receiving Faxes Automatically

The fax machine icon on the right side of the Administrator screen is visible during manual reception and whenever automatic reception is enabled (in the Receive Setup dialog box). While WinFax configures itself

FIGURE 7.5

An active status
message in the
WinFax Administrator
screen

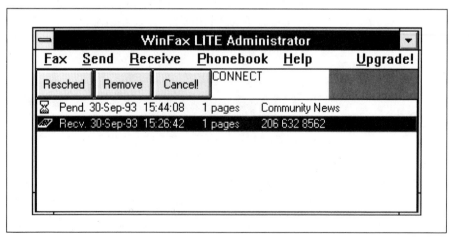

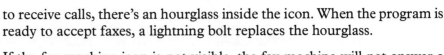

to receive calls, there's an hourglass inside the icon. When the program is ready to accept faxes, a lightning bolt replaces the hourglass.

If the fax machine icon is not visible, the fax machine will not answer a ringing telephone line until you select Manual Receive... from the Receive menu. Chapter 4 contains complete information about receiving faxes.

Recovering Information about Old Faxes

Even if a fax no longer appears on the WinFax Administrator's fax event list, you can still find information about it, and even display, print, or resend it by finding a listing for that event in the Send or Receive Logs. As the names suggest, the Send Log is a list of faxes that you transmitted (and unsuccessful attempts), and the Receive log is a similar list of faxes that WinFax has received. You can select the Send Log from the Send menu, and the Receive Log from the Receive menu.

Working with the Send Log

The Send Log in Figure 7.6 lists outbound fax messages in chronological order. Each listing includes the following information:

- The date and time WinFax sent the fax (or was unable to complete the transmission)
- The name of the recipient (or if you did not specify a name, the telephone number that WinFax dialed)
- The call's status

The status may be either Complete or Error. An error indicates that WinFax failed to send the fax. For more detailed information about an error listing, look at the Status line in the Expanded Information box.

Use your mouse or the up and down arrows to highlight a listing in the log. To use the Fax Viewer to display or print a copy of a fax message listed in the Send Log, click on a highlighted listing, or click on the View... button.

FIGURE 7.6

The WinFax Send Log

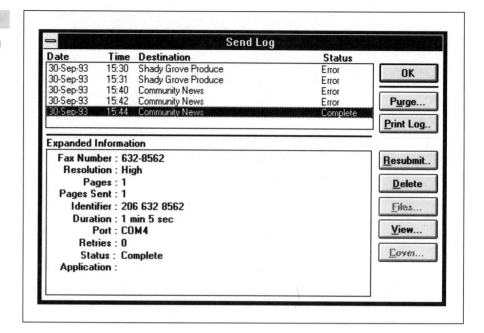

Understanding the Expanded Information Box The Expanded Information box supplies the following information about the highlighted event:

Fax Number	The telephone number that WinFax dialed to originate the call.
Resolution	The resolution setting that WinFax used to transmit this fax. Resolution will be either High (200 x 200 dpi) or Low (100 x 200 dpi).
Pages	The total of number of pages in the document.
Pages Sent	The number of pages that WinFax actually sent successfully.
Identifier	The Station Identifier of the fax machine or modem that received the call.
Duration	The total length of the call, in minutes and seconds.

rry I missed your call. was having out-of-office xperience.

Port	The communications port your computer used to send this fax.
Retries	The number of times WinFax attempted to send this fax.
Status	More specific information than the event listing about the completion of the call. Complete indicates that transmission was successful. Any other entry in this field is a description of the reason the transmission failed.
Application	The Windows application from which the fax was sent. When you originate a fax from WinFax, this field says Direct Send.

Reading a Cover Sheet Sometimes you may want to review the cover sheet that you sent with a fax message, rather than the message itself. Click on the Cover... button to display the View Cover Page dialog box shown in Figure 7.7.

Sending Another Copy of an Old Fax Message You can re-send another copy of any fax event listed in the log by highlighting the event listing and clicking on the Resubmit... button. The Resubmit Event dialog box shown in Figure 7.8 will appear.

FIGURE 7.7

The View Cover Page dialog box

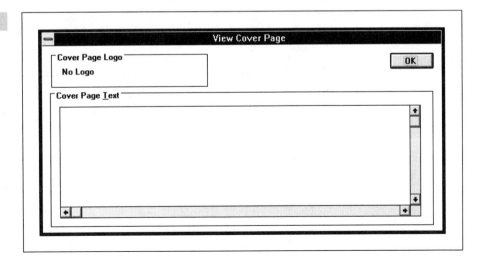

FIGURE 7.8

The Resubmit Event
dialog box

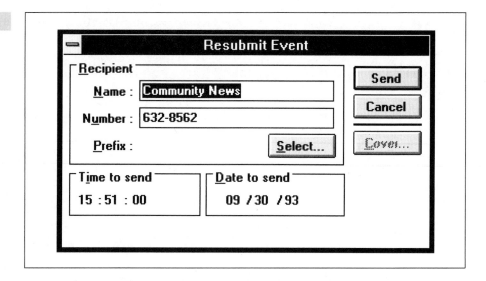

When the Resubmit Event dialog box appears, it shows the name and telephone number that were the destination of the original event and the current date and time. To send another copy to the original recipient or to make another attempt when the earlier transmission failed, click on the Send button.

In some versions of WinFax Lite, this dialog box includes a Send Failed Pages Only option. If this option is available, select it to transmit the pages that didn't make it in the original attempt. To send a copy to a different recipient, either click on the Select button to choose a name and telephone number from your Phonebook, or type over the information in the Name and Number boxes. Click on the Cover... button to edit the original cover sheet.

To schedule a delayed transmission, use the up or down arrows to specify the new date or time in the appropriate box.

When you're satisfied with the new information, click on Send.

Printing the Send Log To print a copy of the Send Log, click on the Print Log... button. Since WinFax uses the Windows Print Manager, it's even possible to send a copy of the log as a fax, if you can think of a reason to do that.

Cleaning Out the Send Log and Deleting Old Images Eventually, some of the listings in your Send Log may outlive their usefulness. When that happens, you can remove individual listings by highlighting them and clicking on the Delete button.

The Purge button gives you three more options:

Purge Fax Data Files	Select this option to delete all the files that hold images of old fax messages. The Send Log will still display information about past fax messages, but you can no longer view or re-send them.
Purge Failed Events	Select this option to remove the records of all fax events with Error status.
Purge All Events	Select this option to remove all fax events from the send log.

Working with the Receive Log

The Receive Log shown in Figure 7.9 lists faxes that WinFax has received through your modem. Each listing includes the following information:

- The date and time WinFax received the fax (or that an attempt failed)
- The station identifier of the fax machine or computer that originated the fax
- The call's status

The status may be either Complete or Error. An error indicates that WinFax failed to receive the complete fax. For more detailed information about an error listing, look at the Status line in the Event Information box.

Use your mouse or the up and down arrows to highlight a listing in the log. To use the Fax Viewer to display a copy of a fax, click on a highlighted listing, or highlight a listing and click on the View button. To print a copy of the highlighted fax message, click on the Print Fax button. To remove the highlighted fax message from the log, click on the Delete button.

FIGURE 7.9

The WinFax Receive
Log

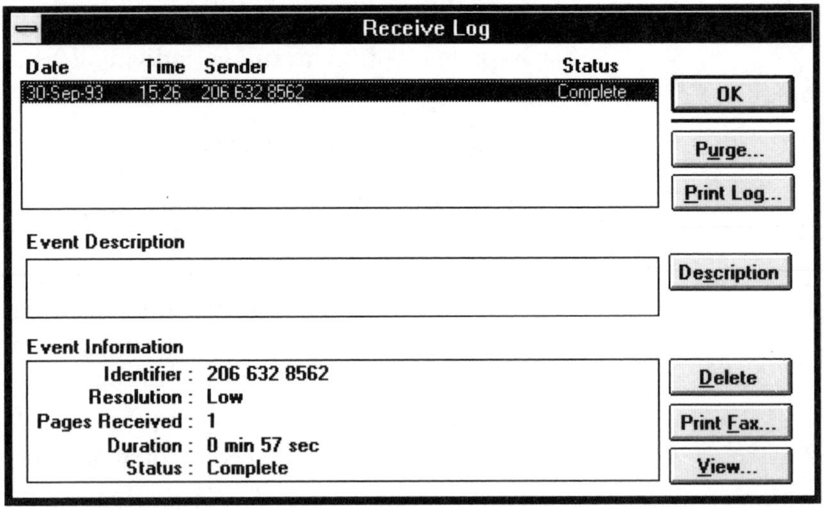

Using the Event Description to Identify a Fax WinFax can attach a brief (80 characters maximum) description to each received fax event, from either the WinFax Administrator screen or the Receive Log. You can use this description to identify the contents of a message in order to make it easier to find in the Receive Log.

If you previously attached a description to a fax event, it appears in the Receive Log's Event Description box. To add, delete, or edit an event description, highlight the fax event listing and click on the Description button.

Understanding the Event Information Box The Event Information box supplies the following information about the highlighted event:

Identifier	The station identifier of the fax machine or computer that originated the call. The identifier is usually either the telephone number or name of the originator of the fax message.
Resolution	The resolution setting that the originator used to transmit this fax. Resolution will be either High (200 x 200 dpi) or Low (100 x 200 dpi).

**Welcome to
the fray!**

Pages Received	The number of pages WinFax received in this message. If the transmission was not complete, this field shows the actual number of pages received, rather than the total number of pages in the transmission.
Duration	The total amount of time that the connection was established between your computer and the other fax machine or other device.
Status	More specific information than the event listing about the completion of the call. Complete indicates that transmission was successful. Any other entry in this field is a description of the reason the transmission failed.

Printing the Receive Log To print a copy of the Receive Log, click on the Print Log... button. As with the Send Log, it is possible to send a copy of the Receive Log as a fax.

Cleaning Out the Receive Log You can clean out old listings from the Receive Log by following the same procedures listed above under "Cleaning Out the Send Log."

Managing Fax Attachments

As we explained in Chapter 3, when you send a fax message, you can attach another file as part of the same transmission. For example, if you're applying for a job by fax, you can write a customized cover letter and add your resume as an attachment.

To display the list of fax attachment files shown in Figure 7.10, select Attachments... in the Administrator's Fax menu.

Select an attachment file from the list to see a description of that file in the Description box. The description includes the original program that created the file and the original file name.

To display the attachment file in the WinFax Image Viewer, highlight the file name and click on the View button.

FIGURE 7.10

The Attachments dialog box

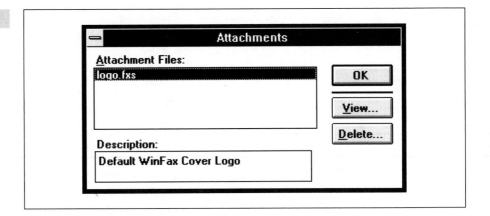

To delete the attachment file, highlight the file name and click on the Delete... button. This erases the .FXS file, but the original source file still remains.

To view or delete more than one attachment at a time, hold down the Shift key and click on each file name you want to highlight.

Changing the WinFax Configuration

To make changes to the original configuration that you specified when you installed WinFax, select the Setup command from the Administrator's Fax menu. Chapter 2 contains information about the WinFax Setup dialog box.

NOTE

WinFax may not save your new configuration settings until you exit the Program. If you normally let WinFax run in background, it's a good idea to close WinFax and restart it after changing the configuration.

Managing Phonebooks

WinFax uses files called *phonebooks* to store lists of frequently called names and fax telephone numbers. You can use a phonebook from the Fax Send dialog box by clicking on the Select button. When you select a name from a phonebook, you automatically instruct WinFax to send a fax message to that person or business without retyping the name or telephone number.

WinFax phonebook files have a .PBK file extension. The default phonebook is PHONE.PBK. Each phonebook can hold up to a thousand listings.

Creating and Editing Phonebook Listings

To add, remove, or change a phonebook record, select Phonebook Record... in the Phonebook menu. The Phonebook Record dialog box shown in Figure 7.11 will appear.

If you're not using the default PHONE.PBK phonebook, type the name of the file you want to use in the Phonebook box or click on the down arrow to display other phonebook files. If Use Partial Matches is active, WinFax will fill in the file name after you type the first few letters.

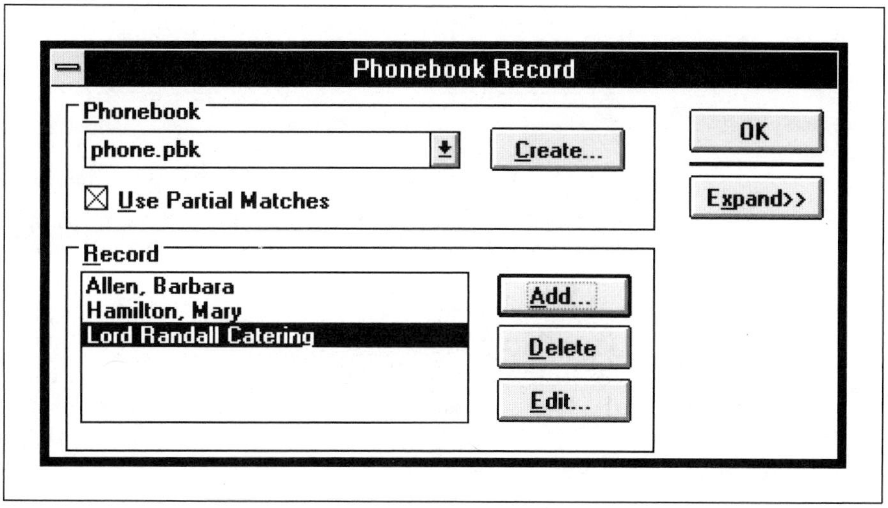

FIGURE 7.11

The Phonebook Record dialog box

To create a new phonebook file, click on the Create... button and type the new name in the Create Phonebook dialog box.

To add, remove, or change a listing in the current phonebook, use your mouse or the arrow keys to highlight that name. To see the complete directory listing for a name, click on the Expand>> button. Figure 7.12 shows an expanded phonebook record.

FIGURE 7.12

An expanded Phonebook record dialog box

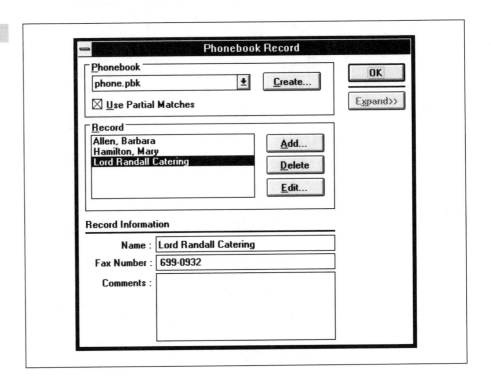

Adding New Names to Your Phonebook

Click on the Add... button to add a new name to the current directory. The Add New Phonebook Record dialog box shown in Figure 7.13 will appear.

Type the name and fax telephone number in the appropriate boxes. If the fax number is a toll call, remember to include the toll prefix (usually *1* in the United States and Canada). It's not necessary to include your standard

```
 ┌──────────────────────────────────────────────┐
 │ ─           Add New Record                    │
 │ Add New Phonebook Record                      │
 │ ─────────────────────────────────────────     │
 │                                               │
 │    Name :  │Sweet Betsey Gold Mining│   ┌─OK─┐ │
 │  Fax Number : │608-555-9301│            └────┘ │
 │  Comments : │Pike County          │  ┌─Cancel─┐│
 │             │                     │  └────────┘│
 │             │                     │            │
 │             └─────────────────────┘            │
 └──────────────────────────────────────────────┘
```

dial prefix (such as *9* for an outside line), since you included it in the Win-Fax Setup dialog box when you installed the program.

Remember that WinFax displays names in a phonebook in alphabetical order. If you're adding a person's name, you may want to type the last name first, so that the directory will list the records that way. Otherwise, you'll end up with all the Freds together in one lump, and all the Susans in another lump.

In the Comments box, you can type a brief description of each entry. For example, you might want to include the voice telephone number for the person or company, or the name of the person to whom you generally send faxes at the company.

Editing an Existing Directory Listing

To change the fax number or comments in an existing record, highlight the name whose record you want to change and click on the Edit button in the Phonebook Record dialog box. When the Edit Phonebook Record dialog box appears, use your mouse or tab key to move to the field you want to change, and type the changes.

You can't edit information in the Name field. To change a name, you have to create a new record.

Removing a Directory Listing

To remove a listing from a phonebook, highlight the name in the Record field, and click on the Delete button. Don't delete a record until you're certain you don't need to save it. It's not possible to undelete a record.

Working with Phonebook Groups

You can send the same fax message to more than one destination by assigning names in a phonebook to a *phonebook group*. When you select a phonebook group from the Fax Send dialog box, WinFax automatically "broadcasts" the fax message to all of the members of the group. Actually, WinFax sends the message to one destination at a time, but it sends an identical fax to each recipient.

You can place the same name in more than one phonebook group, but all the names in a group must be in the same phonebook file.

Creating a New Phonebook Group

To create a new phonebook group, select Phonebook Group... from the Phonebook menu. The Phonebook Groups dialog box shown in Figure 7.14 will appear.

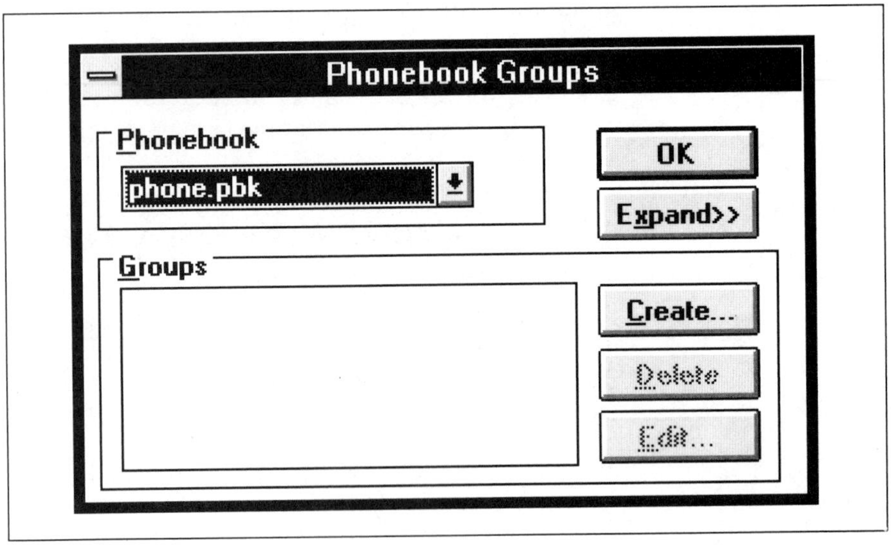

FIGURE 7.14

The Phonebook Groups dialog box

When the Phonebook Group dialog box appears, the current phonebook is the default PHONE.PBK file. If you want to use a different phonebook, use the Phonebook field to select that phonebook file.

Click on the Create... button to display the Create New Phonebook Group dialog box shown in Figure 7.15.

Type the name you want to assign to the new group and click on OK. The Edit Phonebook Group dialog box shown in Figure 7.16 will appear.

Click on the Add... button. The Add Records To Group dialog box shown in Figure 7.17 will appear.

Highlight the first name you want to include in the new group, and click on the Add button. Repeat the process for each additional name you want to include in the group.

When you have added all the names you want included in this group, click on OK.

FIGURE 7.15

The Create New Phonebook dialog box

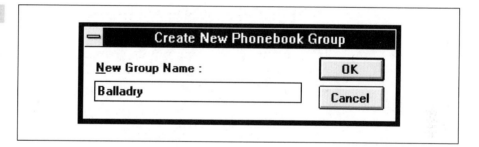

FIGURE 7.16

The Edit Phonebook Group dialog box

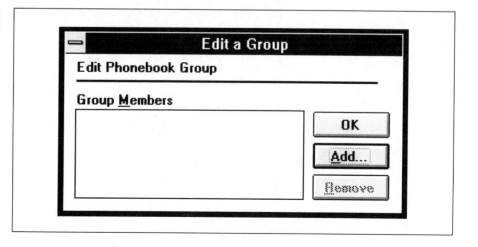

FIGURE 7.17

The Add Records To Group dialog box

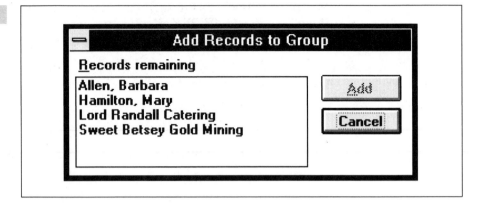

Editing an Existing Phonebook Group

To add or remove names from a phonebook group, or to delete an entire group, select Phonebook Groups from the Phonebook menu to display the Phonebook Groups dialog box shown in Figure 7.18.

To see the current members of a group, click on the Expand>> button. Figure 7.19 shows the expanded dialog box.

Highlight the group you want to change and click on the Edit... button, or double-click on the name of the group. The Edit Phonebook Group dialog box shown in Figure 7.20 will appear.

FIGURE 7.18

The Phonebook Groups dialog box

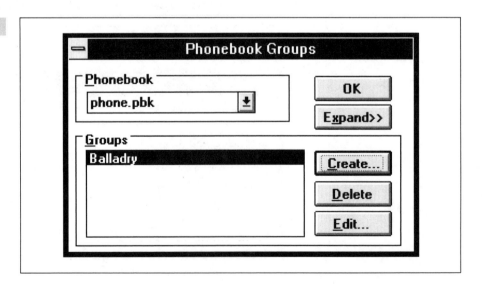

FIGURE 7.19

The expanded
Phonebook Groups
dialog box

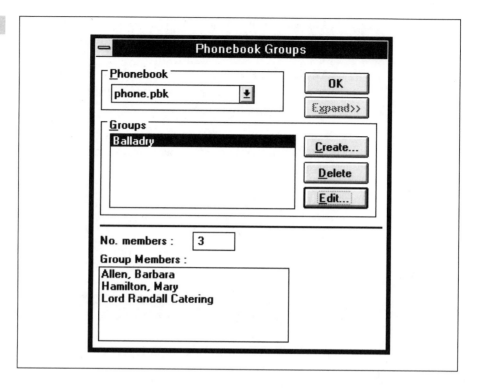

FIGURE 7.20

The Edit Phonebook
Group dialog box

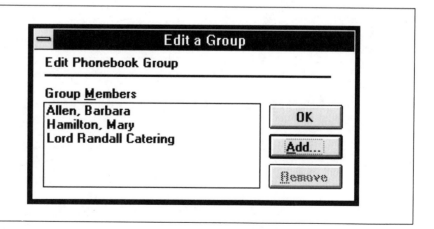

The Edit Phonebook Group shows the current members of the group. Use the ↑ and ↓ keys to see additional names.

Adding Names to a Phonebook Group Click on the Add... button to add new names to the group. Figure 7.21 shows the Add Records to Group dialog box.

Highlight the first name you want to add, and click on the Add button. Repeat the process for each additional name you want to add to the group.

Deleting a Phonebook Group To delete a phonebook group, highlight the name of the group in the Phonebook Groups dialog box, and click on the Delete button.

FIGURE 7.21

The Add Records to Group Remaining dialog box

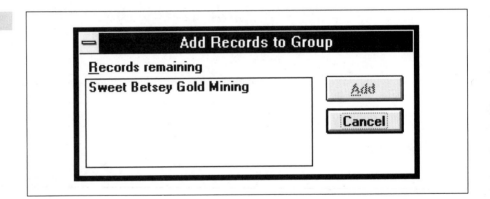

Using a Text File as a Phonebook

When you want to add a lot of names to your phonebook at one time, the Add-type-OK routine can become tedious pretty quickly. As an alternative to typing the names into WinFax dialog boxes one at a time, you can import them to a phonebook as a batch in an ASCII text file. You can also use a database program such as dBASE to create a phonebook file. Of course, the ASCII text file must use the right format, or WinFax won't import it correctly.

You can use the DOS editor, Windows Write, or your word processor to create the list. Remember to save the file as an ASCII text file, without any formatting.

Here's the text format you should use:

"Name","Fax Number","Comments" [carriage return]

Each field must be enclosed in quotation marks and separated by a comma. For example, this file

"Henry, John","1-305-555-1234","Steel Drivers--worked on Big Bend Tunnel contract" ¶
"Hamilton, Mary","1-908-234-5678","Four Marys catering" ¶
"Allen, Barbara","555-6543","Scarlet Town Enterprises" ¶
"Lord Randall","1-206-555-0000","Greenwood Hunters" ¶

produces a phonebook that looks like Figure 7.22.

FIGURE 7.22

A phonebook list taken from an ASCII text file

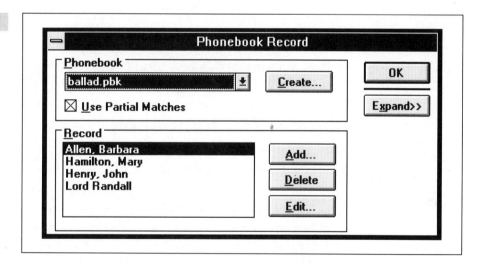

After you create the text file, start the WinFax Administrator program and select Import To Phonebook from the Phonebook menu. Figure 7.23 shows the Import To Phonebook dialog box.

Select the name of the text file from the Import File field. You can either merge the records in the text file into an existing phonebook or create a new phonebook. To make a new phonebook file, click on Create Phonebook…. To merge the text file into an existing phonebook, select the name of the phonebook file from the Phonebook field. When Overwrite Duplications is active, WinFax deletes the old version of any duplicated record.

FIGURE 7.23

The Import To
Phonebook dialog box

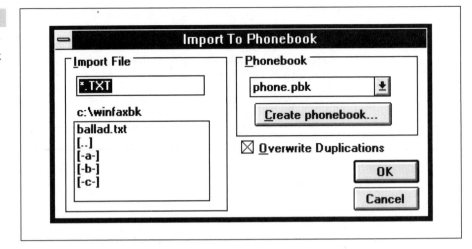

When you click on OK, WinFax will import the text file into the phone-book you specified. When the import is complete, the program displays a message that tells you how many new records were created.

Using a PIM Phonebook

If you're using a Windows PIM (Personal Information Manager), WinFax may automatically recognize the PIM's telephone directory file.

To use a name and fax number from a PIM directory when you prepare to send a fax, click on the Select... button in the Fax Send dialog box. When the Phonebook Entries dialog box appears, open the Phonebook list and select the name of your PIM. WinFax will display names from the PIM directory in the Recipients box.

Other Features of the WinFax Administrator

The WinFax Administrator's menu bar has two more menus that we haven't discussed yet—Help and Upgrade!. The Help menu is similar to those in most other Windows applications, with the usual access to help screens and an About... command that displays information about the WinFax Lite program, including copyright, version number, and release date.

The Upgrade! menu has just one command, WinFax PRO. As a WinFax Lite user, you're entitled to buy WinFax Pro at a discount price. If you decide to order, you can use the WinFax PRO command to create an on-line order form and automatically fax it to Delrina.

PART III

WinFax Pro

WinFax Pro is the retail version of WinFax; in addition to the basic send and receive functions in WinFax Lite, it includes a wide range of advanced features. Among other things, you can import dialing directories from a database, annotate received faxes, and read faxes into text files. Version 4.0 of WinFax Pro adds e-mail, binary file transfer, and auto-forwarding capabilities. In this section, we'll explain all of WinFax Pro's features, with special attention to the differences between Versions 3.0 and 4.0.

CHAPTER

8

Installing
WinFax Pro

AFTER YOU complete the "Setting Up the Modem to Work with WinFax" procedures in Chapter 2, you can be reasonably certain that WinFax Pro will be able to communicate through your modem. This chapter will take you through the WinFax Pro installation process and explain the WinFax settings in the SYSTEM.INI, WIN.INI, and WINFAX.INI files. When you complete these procedures, you'll be ready to learn how to send and receive faxes and use all of WinFax Pro's features.

If you're using an external modem, make sure it is connected to a COM port and turned on before you install WinFax. Otherwise, WinFax won't be able to identify the modem and load the correct initialization string to tell your modem it's sending or receiving a fax message.

During the installation process, you should not be running any other Windows programs or applications except the Program Manager. To make sure there's nothing else running in background, press Ctrl+Esc to use the Windows Task List. If the Task List includes anything besides Program Manager, highlight each of the other tasks and click on the End Task button. When Program Manager is the only remaining task, click on Cancel to close the Task List.

WinFax Pro won't install properly if an earlier version of WinFax has been running during the current Windows session. If you're upgrading from WinFax Lite, follow these steps before you install WinFax Pro:

1. Start Windows.

2. If WinFax Lite starts from the Windows Startup group, close WinFax Lite.

3. Delete WinFax Lite from the Startup group.

4. Close Windows.

5. Restart Windows and install WinFax Pro.

Loading the Program

To install WinFax Pro, start Windows, insert Disk 1 into your floppy disk drive, and select the Run option from the Windows Program Manager. Type **a:\setup** (use the correct drive letter for the drive that contains the diskette) and press ↵. The WinFax Setup dialog box shown in Figure 8.1 will appear.

FIGURE 8.1

The WinFax Pro Setup dialog box

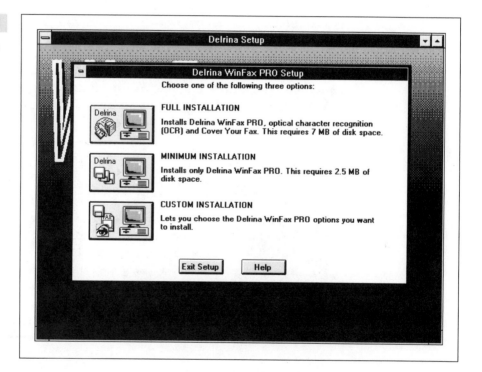

The Setup dialog box offers three options:

Full Installation   This option loads the WinFax program files, the OCR program that converts fax images to text files, the library of cover pages, marco that link WinFax to applications, and support for electronic mail.

Full installation requires at least 4.5 megabytes of extended memory and 7 megabytes of free space on your hard drive. If you don't have that much memory or disk space, or if you don't plan to use all the optional features, choose one of the other options.

Minimum Installation   This option loads the WinFax program files only. Minimum installation requires at least 2 megabytes of extended memory and 3 megabytes of free space on your hard drive.

Custom Installation   This option allows you to choose the programs you want to install. Choose this option to load some optional features, but not all of them. You can also use Custom Installation to exclude one or more group of cover pages.

If you choose minimum or custom installation, you can install additional options later by running the WinFax Setup program again from Disk 1. When the program asks if you want an update or a new install, click on the Update button.

TIP

The cover page library requires about 1.6 megabytes of disk space. You can find pictures of all the cover sheets in the library in the *Cover Your Fax* booklet included in the WinFax package. You probably won't want to use all of them. Unless you have unlimited space on your hard drive, you should consider choosing Custom Installation and using the Options button in the Installation Options dialog box to exclude the cover pages groups that you'll never use.

Continue through the setup process until the dialog box in Figure 8.2 appears. Select the port that is connected to your modem. If the COM port connected to your modem is not one of the available ports in the dialog box, make sure the modem is turned on and click on the Re-Test button. If WinFax still can't find the modem port, click on Exit Setup and make the changes described in Chapter 2, under "Configuring Windows for an Internal Modem."

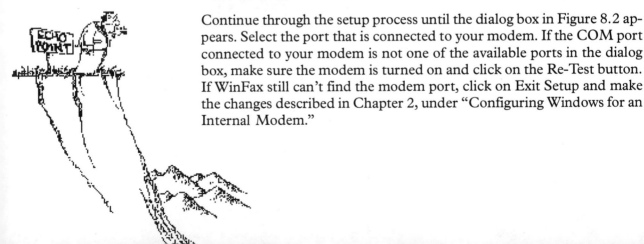

FIGURE 8.2

The WinFax Setup Port
Selection dialog box

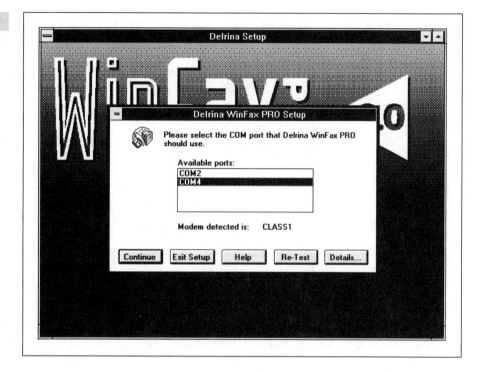

When WinFax lists your modem port in the list of available ports, highlight that port and click on the Continue button. WinFax will display the list of modems shown in Figure 8.3

In most cases, WinFax will automatically highlight the brand and model of your modem. If it does not, use the arrow keys to search for your modem type and highlight it. Notice that the Modem Init String changes for each modem type.

If you can't find your modem in the list, go back to the top of the list and select one of the generic types. Look in the specifications section of your modem manual to find out if the modem type is Class 1 or Class 2, and highlight that generic description.

Class 1 and Class 2 are international standards for facsimile transmission. In Class 1, the computer manages the communications session, while in Class 2, the firmware in the modem handles communication. Therefore, a Class 2 modem moves some of the processing burden away from the

FIGURE 8.3

The WinFax Setup
Modem Selection
dialog box

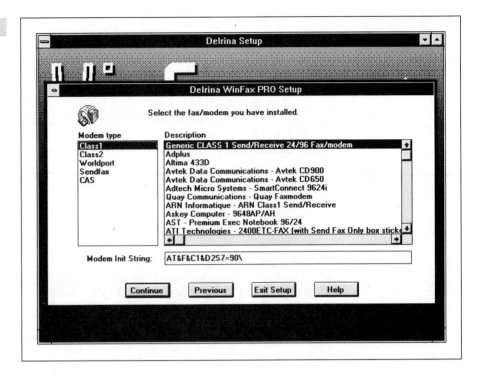

computer and onto the modem. In WinFax Pro 3.0, it doesn't matter which one you choose as long as WinFax knows what kind of modem it's talking to; it works equally well with Class 1 and Class 2. In WinFax Pro 4.0, use Class 1 if possible, since Error Connecting Mode doesn't work with Class 2.

The Intel SatisFaxtion modem uses a third standard, called CAS (Communications Application Specification). CAS controls some of the stuff automatically that WinFax controls for other modems. As a result, WinFax uses a handful of slightly different screens with a CAS modem.

If you're using a no-name fax modem that you bought at a swap meet and WinFax doesn't identify it, the best you can do is to try one of the generic listings. Start with Class 1, and if that doesn't work, try again with Class 2. If neither setting works, look for an initialization string in the modem manual (you *did* get a manual with that thing, didn't you?) and type it into the Modem Init String field. If that fails, you might want to seriously consider using the modem as a paperweight.

After you highlight your modem type, click on the Continue button. At this point, WinFax will start to copy files from the diskettes to your hard drive. Follow the instructions that appear on your screen.

When WinFax asks you to define WinFax as the default printer, choose No unless you expect to send faxes more often than you print things. If you share your computer with other users who will (quite reasonably) expect to print when they enter a Print command, choose No. If you do want WinFax as the default, choose Yes. You can always change the default printer later from the Windows Print Manager if necessary.

WinFax will now display a series of setup dialog boxes and wait for you to fill them in before moving to the next dialog box. We'll explain each setup dialog box in the following sections.

The Program Setup Dialog Box

The Program Setup dialog box is the place where you specify the information that WinFax automatically places on each page it sends, the station identifier it sends when it connects with another fax machine or modem, and some of the program's user interface features. Figure 8.4 is a sample Program Setup dialog box.

FIGURE 8.4

The Program Setup
dialog box

Program Setup

Header
Left : `From: @S  To: @R`
Center : `Date: @D  Time: @T`
Right : `Page @P of @N`

[Insert **V**ariable...]   [Defa**u**lt]

General
Sta**t**ion Identifier (CSID): `John Ross in Seattle`

[X] Displa**y** Call Progress
[X] Use **P**artial Matches in Phonebook
[ ] Pro**m**pt for Billing Code/Keyword
[ ] Keep Only Active **W**indow Open
[ ] **S**ave Window States on Exit
[ ] Save W**i**ndow Sizes on Exit

[OK]
[Cancel]

[**L**og...]
[Win**F**ax Driver...]

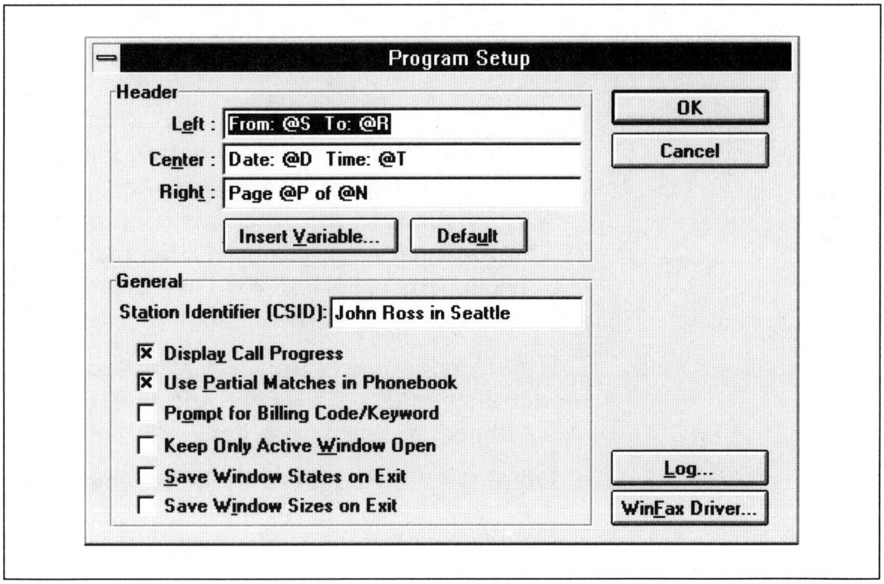

The Program Setup dialog box includes these options:

Header   The header is the line of small print across the top of each page, which WinFax adds to faxes as it sends them. In the United States, there's a federal regulation that either the cover sheet or the header must identify the date and time the message was sent, the name *or* company name of the sender, and the sender's telephone or fax number.

The Left, Center, and Right fields correspond to the location in the header where the text string in each field will appear. To automatically place current information in a header, place the cursor in a field and click on the Insert Variable... button. Choose the variable item you want to use, and click on OK. The header fields shown in Figure 8.4 will produce a header that looks like Figure 8.5.

FIGURE 8.5

A typical fax header

FACSIMILE COVER PAGE

To: From:
Time: 13:04:24   Date: 11/23/93
Pages (including cover):   1

This is a sample WinFax Lite Cover Page.

Station Identifier   The station identifier (sometimes called the CSID, for *calling station identifier*) is a string of characters that WinFax sends to the other fax machine or modem to identify you. This information appears on a fax machine's digital readout and in fax activity reports. Type your name, your company name, or your fax number in this field.

If your modem uses the CAS standard, you won't see the Station Identifier field, because you specified the station identifier in the CAS modem setup program.

Display Call Progress   When this option is active, WinFax displays a pop-up status message whenever it sends or receives a fax. If you don't want to see these messages, click on this option to remove the check mark.

Use Partial Matches in Phonebook   When this option is active, WinFax will search for names and fax numbers in your phonebook when you start to type a name in the Send Fax dialog

box. For example, if you type **all**, WinFax will find "Barbara Allen" in your phonebook, and fill in the rest of the name and the fax number automatically.

Prompt for Billing Code/Keyboard (Version 3 only)   When this option is active, WinFax will ask for a keyword and billing code every time you send a fax. You can include this information in cover sheets and in your transmitted calls logs.

Show Command Icon Text (Version 3 only)   When this option is active, there is a label under each command icon in the buttons on the WinFax Pro screen.

Keep Only Active Window Open (Version 4 only)   When this option is active, WinFax automatically closes the last window you used when you open a second window. If this option is not active, old windows will remain open.

Save Window States on Exit (Version 4 only)   When this option is active, all of the windows that are open when you exit WinFax will remain open the next time you start the program.

Save Window Sizes on Exit (Version 4 only)   When this option is active, each open window will be the same size the next time you start WinFax.

Log... (Version 4 only)   In Version 4 of WinFax, you can use the Log Setup dialog box to automatically delete events or pages after a specified number of days. Click on the Log... button to display the Log Setup dialog box.

WinFax Driver... (Version 4 only)   The Driver Setup dialog box in Version 4 is identical to the Setup dialog box you will get when you click on Printer Setup in an application. It specifies the page size and orientation, and the resolution (in dots per inch) of fax images transmitted by WinFax.

The Dial Setup Dialog Box

The WinFax Pro 3.0 Dial Setup dialog box controls the way WinFax operates your modem and contains your telephone credit card number and area code. Figure 8.6 shows the Dial Setup dialog box. In WinFax Pro 4.0, these options are in the Fax/Modem and User dialog boxes.

FIGURE 8.6

The Dial Setup dialog box

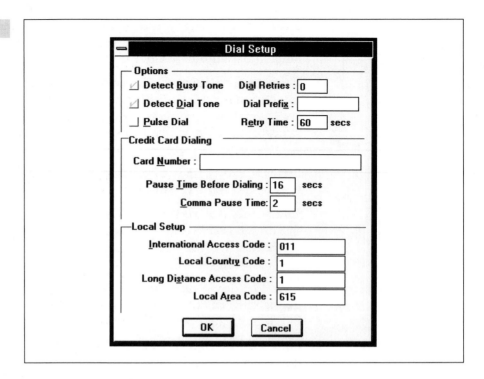

The Dial Setup dialog box includes these options:

Detect Busy Tone   When this option is active, your modem will automatically hang up when it detects a busy signal. WinFax will try to place the call again the number of times specified in the Dial Retries field. If the line is still busy, WinFax will log the attempt with an error message. As a general rule, you should make this option active.

Detect Dial Tone   When this option is active, your modem will listen for a dial tone before it starts to dial. Unless you're using WinFax with some kind of unusual PBX system, this option should be active.

Pulse Dial   When this option is active, your modem will use dial pulses instead of tones to dial outgoing calls. Do not make this option active unless your telephone service does not support pushbutton Touch-Tone dialing.

Dial Retries   This option specifies the number of times Win-Fax will try to place a call after it detects a busy signal. The Retry Time field sets amount of time between retries.

Dial Prefix   This option automatically adds a prefix to each call. If you have to dial *9* (or some other number) for an outside line, that's your dial prefix. If you have to use a different prefix for a WATS line or a tie line, you can change the dial prefix from the Send Fax dialog box when you place a call.

Retry Time   This option sets the number of seconds WinFax waits between retries when it detects a busy signal.

Credit Card Dialing   If you use a telephone credit card to send faxes, type your credit card number in the Card Number field. The dialog box displays asterisks in place of your credit card number. WinFax will dial your credit card number when you se-lect Use Credit Card in the Send Fax dialog box.

The *Pause Time Before Dialing* field sets the number of seconds WinFax waits for the second dial tone before it starts to dial the credit card number. Don't change the default setting.

The *Comma Pause Time* field sets the number of seconds WinFax pauses when there's a comma in the dial string. Set this option to 10 seconds.

The *Dial Prefix* field specifies the telephone number your long dis-tance carrier uses for dial-in credit card access. If you use a toll-free 800 number to make calls on your credit card, type the number in this field, and place a comma at the end. Otherwise, leave this field blank.

The *Local Setup box* includes these options:

International Access Code   The international access code is the prefix you must dial for international calls. In North America, use the default *011*. In other parts of the world, look in your local telephone directory for the correct code.

Local Country Code   The local country code is the prefix you must dial to place a long distance call within your own coun-try or region. In North America, the access code for calls within the United States and Canada is *1*.

Long-Distance Access Code   The Long-Distance Access Code is the number that you dial before you dial a long-distance number. In most parts of North America the standard access code is 1. If you're not sure about your long-distance access code, look in the front of your local telephone directory.

If you use a different long-distance access code to place a call through a PBX or Centrex system, ask your telephone system manager for the correct long-distance access code.

Local Area Code   The local area code is the area code that people in other area codes use to call you. If you place a call to your own area code, WinFax will not dial the area code.

If you're using WinFax on a portable computer, remember to change this field when you travel to a new area code.

NOTE

In some parts of the United States, you must dial the area code when you want to place a long-distance call within the same area code. If you are calling from one of those areas, type a dash (-) before the area code when you enter the fax number in your phonebook or the Send Fax dialog box. For example, if you're in Seattle and you want to send a fax to 555-5839 in Bellingham, Washington, you must type the Bellingham fax number as -206-555-5839.

The User Setup Dialog Box

The User Setup dialog box contains information that WinFax will use in variable fields on cover sheets. Figure 8.7 shows the User Setup dialog box.

Type your name, your company name, fax number, and voice telephone number in their respective fields. Use the exact text string you want to appear on your cover sheets.

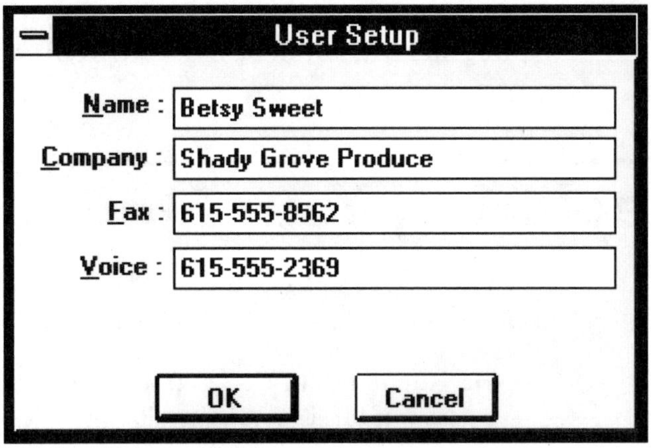

FIGURE 8.7

The User Setup
dialog box

Upgrading

When you install WinFax Pro as an upgrade from WinFax Lite, or from an earlier version of WinFax Pro, the setup program will offer you the chance to save the phonebooks you used with the earlier version of the program and to extract information for the setup dialog boxes from your existing configuration.

Before you try to use the upgraded version of WinFax, exit and restart Windows to make sure the changes take effect.

Completing Your Installation

After you complete the setup dialog boxes, WinFax may offer some changes to your SYSTEM.INI file. These changes will improve reliability at the expense of communication speed. If the program recommends that you make these changes, go ahead and accept them.

Finally, WinFax will offer you the chance to test your installation by faxing your software registration to Delrina. If you choose No in Version 3, the Register Menu will be available from the WinFax screen.

Replacing Your Modem

Follow these steps if you install a different kind of modem after you install WinFax, or if you specified the wrong modem during installation:

1. Run the WinFax Setup program from Disk 1. When the program asks if this is a new installation, click on the Change Modem button.

2. Highlight the COM port connected to your modem and click on the Re-Test button.

3. When the Modem Select dialog box appears, highlight the description of your new modem and click on the Continue button.

Configuring Your .INI Files for WinFax

When you install WinFax, the Setup program automatically adds a [WIN-FAX] section to the WIN.INI file and creates a new WINFAX.INI file. If necessary, the program may also make some changes to the SYSTEM.INI file.

For the most part, you can make any necessary changes to these files by changing WinFax dialog boxes. If WinFax is working properly, don't mess with the .INI files. There are, however, a handful of items that you can only change directly in the .INI files:

OCR In Low Memory (version 3 only)   It's not the best way to go, but it's possible to run the OCR program in low memory. If you choose to run OCR in low memory, add this line to the [Any-Fax] section of WIN.INI:

TempDiskSpace=n

In place of n, specify the number of kilobytes of memory you want to allocate to OCR. The default is 1000.

Do Not Display Thumbnails   Normally, the Cover Page Library dialog box displays small pictures of each cover page file, but this can take a lot of time on a 386SX or other PC with limited resources. To disable thumbnail displays, add this line to the [General] section of the WINFAX.INI file:

ShowThumbnails=No

Do Not Compress Archives   WinFax normally stores fax files in compressed form, in order to reduce the amount of disk space they occupy. If you turn off compression, it will take less time to display them, but they will take up more space. To disable compression, add this line to the [General] section of the WIN-FAX.INI file:

Compress=0

Change the Credit Card Access Code   In most parts of North America, the dialing prefixes for credit card calls are 0+ for calls to the U.S. and Canada, and 01+ for calls to other countries. WinFax uses these codes as the default. If your telephone system uses different prefixes for operator-assisted direct-dialed calls, add these lines to the [General] section of the WINFAX.INI file:

CreditCardLongAccess=[prefix]
CreditCardIntlAccess=[prefix]

Change the Fax Vacuum Threshold   The WinFax Viewer's Cleanup Fax command removes random black specks on faxes received through noisy telephone lines. You can increase or decrease the program's ability to identify and remove these spots by adding this line to the [Viewer] section of the WINFAX.INI file:

FilterThreshold=n

n may be any value from 2 to 6. Set the threshold to 2 for the least amount of cleanup. The default value for *n* is 5.

If you delete the WINFAX.INI file, WinFax will generate a new file the next time you run the program. However, the new file won't include the [WinFax Importers] section. To restore that section, use Windows Write or the DOS Editor to copy it from the README file in your WINFAX directory to the new WINFAX.INI file.

Starting the WinFax Programs

When installation is complete, you will have a new group called Delrina WinFax PRO in the Windows Program manager, with three programs and two information files. Figure 8.8 shows a Delrina WinFax PRO group window. Yours may be sightly different.

FIGURE 8.8

The WinFax PRO Group window in the Windows Program Manager

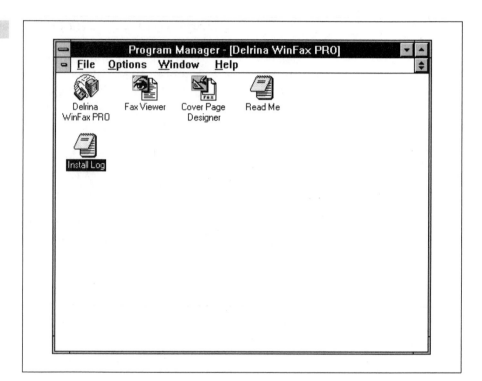

To start the main WinFax program, double-click on the Delrina WinFax PRO icon. We'll explain all the things you can do from this program in the next few chapters.

If you want WinFax to automatically receive faxes whenever Windows is running, you should configure the program to start automatically. In Windows 3.1, drag the Delrina WinFax PRO icon from the WinFax PRO group to the StartUp group in the Windows Program Manager. In Windows 3.0, add this line to the [Windows] section of the WIN.INI file:

LOAD=FAXMNG.EXE

If you're using some other Windows desktop instead of the Program Manager, you may need to know that the main WinFax program is called FAXMNG.EXE.

To start the Fax Viewer or Cover Page Designer without opening the main WinFax application, double-click on that icon. If the main WinFax program is already running, you can start the Fax Viewer and Cover Page Designer from the menu or the icon ribbon.

The Read Me file contains information about the latest revision to WinFax and a backup copy of the [WinFax Importers] section of the WINFAX.INI file. Double-click on the Read Me icon to read this file.

The Install Log file is a text file called WINFAX.LOG that contains information that you specified when you installed WinFax. If you ask Delrina's Product Support group for help, they may ask you for information from this file. To read this file, double-click on the Install Log icon.

9

Receiving Faxes with WinFax Pro

YOU CAN decide when to send a fax, but incoming faxes arrive at your computer without warning. And like a telephone ringing while you're in the bathtub, faxes are likely to arrive while you're trying to do something else with your computer. You could set up a second computer to do nothing but receive faxes, but then what's the point of using a computer to receive faxes in the first place? You might just as well have a stand-alone fax machine. WinFax has a better solution to this problem: it offers you the option of receiving faxes in the background, both in Windows and through a memory-resident DOS program.

This chapter explains how to set up WinFax to receive faxes in both DOS and Windows and describes the process of receiving a fax. Chapter 12, "Viewing and Printing Fax Messages," will tell you how to use the Fax Viewer program to display faxes on your screen and print paper copies.

Configuring WinFax to Receive in Windows

When you installed WinFax Pro, the program was configured to answer incoming calls automatically and receive faxes whenever WinFax is running and your modem is on. If you're using the same telephone line for both voice and fax calls, auto-answer may not be the best option, because your voice callers will hear a high-pitched tone (known to telephone engineers as a CNG tone) when the modem answers.

To change the default receive configuration, in WinFax Pro 3.0 open the Receive Setup dialog box from the Receive menu on the main WinFax screen. In WinFax Pro 4.0, use the Receive command in the Setup Menu. Figure 9.1 shows the default Receive settings.

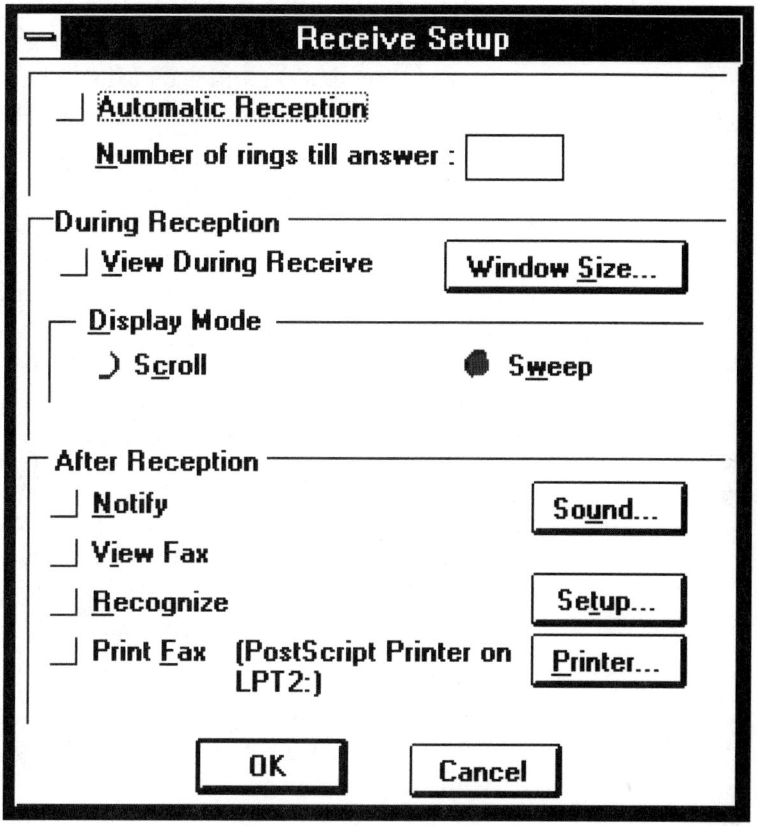

FIGURE 9.1

The WinFax Pro 3.0 Receive Setup dialog box with default settings active. The 4.0 dialog box is similar.

You can choose from the following settings in the Receive Setup dialog box:

Automatic Reception   When Automatic Reception is active, WinFax will answer your telephone after it rings the number of times specified in the Number Of Rings Till Answer field. When your modem hears the CNG tone from the calling fax machine or modem, it exchanges handshake signals and starts receiving.

If you have a dedicated telephone line for fax calls, you should make Automatic Reception active and set the number of rings to 1. If you use the same telephone line for both fax and voice, you have two options. If you're using WinFax in an office setting, where there are no extension telephones on the same line in other locations, make Automatic Reception active, set the number of rings to 10 (allowing ten rings per minute), and plug your telephone set into the modem's TEL or PHONE jack. When the telephone rings, pick up the handset and listen. If you hear a CNG tone, select the Manual Receive option from the WinFax Receive menu.

Your other option is to deactivate Automatic Reception so your modem will not answer incoming calls until you instruct it to do so. This is a good choice for receiving faxes on a home computer, where there are extension telephones in other rooms. When the telephone rings, answer it on any extension and listen for a CNG tone. If you hear the tone, go to the computer and select the Manual Receive option from the WinFax Receive menu.

The current Automatic Reception status is always visible in the bottom of the Main WinFax screen.

Number Of Rings Till Answer   When Automatic Reception is active, this field specifies the number of times the telephone will ring before your modem answers. The default is one ring.

During Reception   This section of the Receive Setup dialog box instructs WinFax to display incoming faxes while it is receiving them. This section is not available with a CAS modem.

View During Receive   When View During Receive is active, WinFax pops up a window that displays faxes as they arrive at your computer.

Window Size   Use this option to set the size and location of the window that will contain incoming faxes. When you're satisfied with the position of the empty window, close it. It will open automatically when WinFax receives a fax.

Display Mode   This option specifies the way WinFax will display incoming faxes when View During Receive is active. When Scroll is active, WinFax displays faxes starting from the bottom; when Sweep is active, WinFax starts at the top of the window and works down. Since faxes scan from the top, that's the way they're transmitted. Therefore, WinFax has to wait until it receives the entire page before it starts to "scroll," but it displays top-down "sweeps" as it receives them. If you make View During Receive active, choose Sweep mode.

After Reception   This section of the Receive Setup dialog box controls the way WinFax handles faxes after it receives them.

Notify   When Notify is active, WinFax sounds a signal and displays the message in Figure 9.2 when it has received a new fax message. The number in the message increases every time WinFax receives another fax. To reset the number to zero, click on OK.

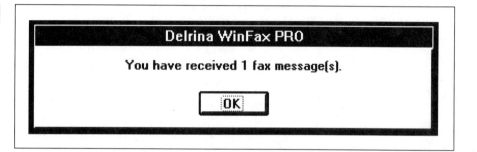

Sound   Click on the Sound button to choose the audible signal that WinFax will use when it receives a fax from the dialog box in Figure 9.3.

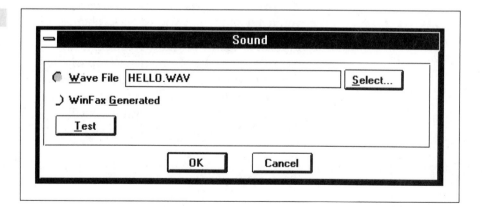

If you have Windows 3.1 and a sound board, choose Wave File to use a Windows .WAV file as your signal. Use the Select button to change the file. You can use any of the standard Windows .WAV files, such as bells or chimes, or you can create your own file. You could even use a microphone or voice synthesizer to make your computer say something like, "I've just received a fax. Would you like to see it now?" every time a new fax comes in.

If you don't have a sound board, choose WinFax Generated to use the default beeps.

To listen to the signal you've specified, click on the Test button.

View Fax   When View Fax is active, WinFax will automatically start the Viewer program and display the first page of the new fax. Chapter 12 contains detailed information about using the Viewer program.

Recognize   When Recognize is active, WinFax automatically uses the OCR program to convert faxes to text files as it receives them. This option will only work if you loaded the OCR program when you installed WinFax. Chapter 15 explains the OCR program in more detail.

Like the Print Fax option, automatic recognition ties up WinFax so that it can't receive another fax while it is converting. And of course, if you receive a graphic by fax, the OCR program won't be able to convert it. So in most cases, it's probably not a good idea to automatically recognize every fax as it comes in. You can convert the ones you need later, after you've had a chance to read them.

If you do decide to make Recognize active, click on the Setup button to display the Recognize Setup dialog box in Figure 9.4. Choose Store Text In Receive Log or Save Text To File to tell WinFax where to store the converted texts of received faxes. If you select Save Text To File, a Path field will appear in which you can specify the directory where you want to store text files.

The File Format field defines the way the program will store text files. If you plan to import the files to a word processor, use ASCII Text or ASCII Text with CRs. CRs are carriage returns at the end of each line in the original fax document. If your word processor supports it, you might preserve some of the text formatting in the original by storing the text file in Microsoft RTF format. To export the fax to an Excel document, use Excel Text format. To display the text file on your screen, use one of the ANSI Text formats.

FIGURE 9.4

The Recognize Setup
dialog box

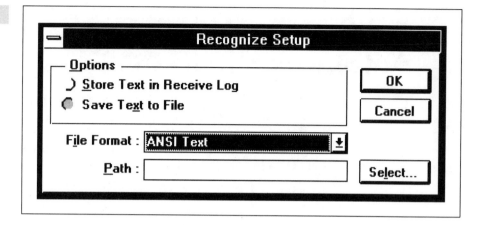

Print Fax   When Print Fax is active, WinFax prints faxes immediately after receiving them, using the printer named in the Receive Setup dialog box. Click on the Printer button to change printers.

WinFax can't receive another fax while it is automatically printing. Since you really can't predict when people will send you faxes, it is a good idea to make this option inactive.

In WinFax Pro 4.0, the Receive Setup dialog box includes two additional options. When Forward is active, WinFax automatically re-sends all faxes that it receives. The Remote Retrieval option allows callers to call your fax modem and receive a specified file from your system. For more detailed information about both of these new features, see Chapter 16, "New Features in WinFax Pro 4.0."

If you configure WinFax to answer automatically, you should also set up the Windows Program Manager to start WinFax every time you start Windows. Automatic startup is a simple process: open the Startup group and the WinFax Pro group in the Windows Program Manager, and use your mouse to drag the WinFax Pro icon to the Startup group. Since you won't need to do anything else in the Startup window, you can minimize it to reduce the amount of clutter on the Program manager screen.

The next time you start Windows, WinFax will start automatically. Click on the down arrow in the upper-right corner to minimize the WinFax screen.

Configuring WinFax to Answer Automatically in DOS

Even though WinFax is a Windows program, it includes a memory-resident DOS utility to receive faxes when Windows is not running and when DOS is running within a full-screen window. You can load this program by hand when you close Windows and return to DOS, or you can start it from your AUTOEXEC.BAT file every time you turn on your computer.

To start the DOS TSR (Terminate and Stay Resident) program, enter this command at the DOS prompt, or place it in your AUTOEXEC.BAT file:

    WFXTSR -W*path*

Use the path of the WinFax directory in place of *path*. For example, if you used the default WinFax directory, this would be the command:

    WFXTSR -WC:\WINFAX

If you're in the WINFAX directory already, you can leave out the -W*path*.

The DOS program uses about 33K of memory. If you place the program in your AUTOEXEC.BAT file, run the MS-DOS MemMaker program or your memory manager's optimization program to move the WFXTSR program to high memory after you load it.

In order to automatically receive faxes in DOS, the WFXTSR program must be resident in memory, and the WinFax program within Windows must be configured for Automatic Reception. If Automatic Reception is not active in Windows, you must receive calls manually in DOS.

To remove the fax program from your computer's memory, enter this command:

    WFXTSR -U

Manually Receiving Faxes

If you're using the same telephone line for both fax and voice, you may not want WinFax to answer calls automatically. When you disable automatic reception, WinFax does not answer incoming calls until you enter a "Manual Receive" command. You can take a fax call manually from both Windows and DOS.

Receiving Manually in Windows

To turn off Automatic Reception in Windows, select the Receive Setup command from the Receive menu in WinFax 3.0, or the setup Menu in WinFax Pro 4.0. When the dialog box in Figure 9.1 appears, click on Automatic Reception to remove the check mark. Then click on OK to close the dialog box.

When a call comes in, follow these steps to receive a fax manually:

1. When the telephone rings, pick up the handset, say "hello" or whatever you usually say when you answer the telephone, and listen. If somebody speaks to you, this is a voice call. If you hear a high-pitched CNG tone, it's a fax call.

2. Start WinFax (or maximize it if it's minimized), and select the Manual Receive command from the Receive menu in WinFax Pro 3.0 or the Manual Receive Now command in WinFax Pro 4.0. Don't hang up the handset yet.

3. When WinFax asks if you want to answer the call now, click on the Yes button. The current status of the call will appear in the Status Message Area at the bottom of the WinFax screen.

4. After WinFax answers, hang up the handset.

Receiving Manually in DOS

To keep WinFax from automatically answering calls in DOS, you must turn off Automatic Reception in Windows. Start Windows and WinFax, select the Receive Setup command from the Receive menu, and click on Automatic Reception to remove the check mark. Then return to DOS by closing the dialog box, closing WinFax, and exiting Windows.

When Automatic Reception is not active and WFXTSR is resident in memory, follow these steps to receive a fax from DOS manually:

1. When the telephone rings, pick up the handset, say "hello," and listen. If somebody speaks to you, this is a voice call. If you hear a CNG tone, it's a fax call.

2. To receive a fax, hold down the Shift key and the Ctrl key, and press ↵.

3. If Display Call Progress is active in the WinFax Program Setup dialog box, the status of the current call appears in the upper-right corner of the screen, as shown in Figure 9.5. If Display Call Progress is not active, the WinFax TSR will receive and store the incoming fax, but it won't show the call's status on your screen.

FIGURE 9.5

The WinFax DOS TSR program shows current call status

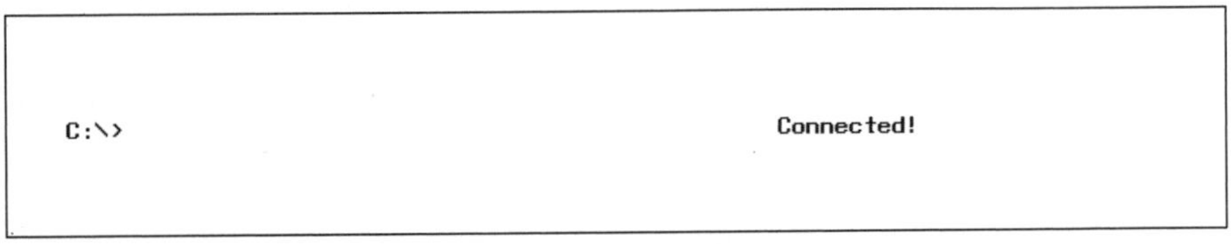

```
c:\>                                          Connected!
```

To view or print a fax that you received while in DOS, you must start Windows and the WinFax Viewer. It's not possible to display a fax from the DOS prompt.

Congratulations!

Welcome to the fray!

Working with WinFax Pro Phonebooks

WINFAX USES phonebooks to store lists of frequently-called names and fax numbers. When you place the name of a person or business to whom you send faxes in a phonebook, you can select it from a list or just type the first few letters of the name, and Win-Fax will automatically fill in the recipient's fax telephone number. You can also assign names in a phonebook to group and send the same fax message to all the members of a group with a single command. If you're using a database program that can create lists in dBASE format, it's possible to transfer lists directly from your database to a WinFax phonebook.

This chapter contains complete instructions for creating phonebooks and for moving names and telephone numbers between files created by other applications and WinFax phonebooks. In the next chapter, we'll tell you how to send faxes to individual phonebook listings and to groups. Phonebooks in versions 3.0 and 4.0 of WinFax Pro use very different screens, so we'll describe the two versions separately.

A WinFax Pro phonebook can contain any number of entries, so it's quite possible to put all of your names and numbers in the default phonebook. But you might also decide to create more than one phonebook, each with a different listing category. For example, you might want to keep your list of customers in one phonebook and suppliers in a separate list. Keeping different classes of records in separate phonebooks has real advantages if you plan to print copies of phonebooks and use them for other purposes.

Phonebooks in WinFax Pro 3.0

The WinFax 3.0 Administrator screen includes a Phonebook menu with five commands. You can use the PhnBook icon button instead of the Records... command to open the Edit Phonebook dialog box.

Creating a New Phonebook

To create a new phonebook, select the Phonebook List… option from the Phonebook menu in the main WinFax screen. When the Phonebook List dialog box shown in Figure 10.1 appears, click on New….

FIGURE 10.1

The Phonebook List dialog box

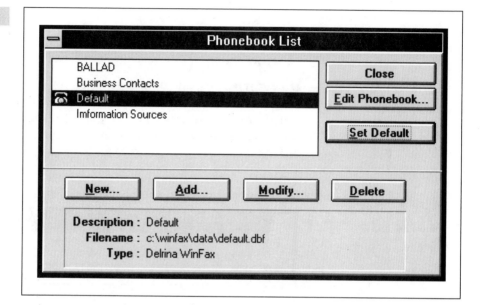

The New Phonebook dialog box shown in Figure 10.2 has three sections:

Description   Use the Description box to describe the new phonebook (up to 40 characters). Unlike most computer lists, you don't have to condense a complicated description like "Press Release Distribution List" into some kind of cryptic abbreviation that you won't recognize three weeks later. You can actually use a description that makes sense (What a concept!).

Filename Prefix   Of course, you can't completely escape those cryptic abbreviations. Type a filename of up to seven characters in this box. WinFax will use this name for the seven files it uses for each phonebook, with different final characters and file

extensions. For example, if you assign a filename prefix of
PRESREL, WinFax will create these files:

PRESSREL.DBF

PRESSRELG.DBF

PRESSREL1.NDX

PRESSREL2.NDX

PRESSREL5.NDX

PRESSREL6.NDX

PRESSREL7.NDX

Directories   The default location for phonebook files is the
DATA subdirectory in the directory that contains the WinFax
files. In most cases, that's a fine place to store them. But if you
want to put this phonebook in another directory or on a floppy
disk, you can use this command.

FIGURE 10.2

The New Phonebook
dialog box

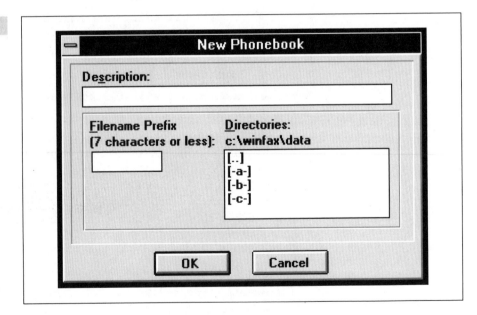

After you enter a description and filename, click on OK to return to the
Phonebook List dialog box. The new description should now appear in
the list. Double-click on the description to immediately edit the new
phonebook.

Adding Records to a Phonebook

Use the Edit Phonebook dialog box shown in Figure 10.3 to make changes to an existing phonebook. There are three ways to display the Edit Phonebook command:

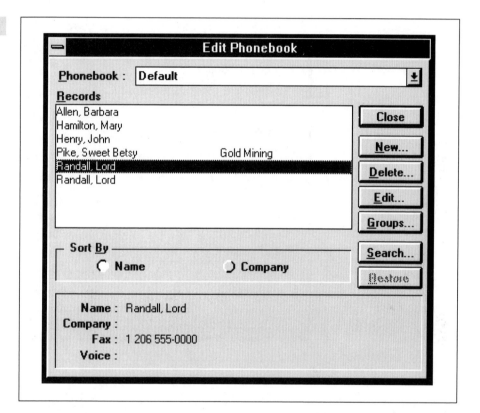

- Click on the PhnBook icon
- Select the **Records** command in the Phonebook menu
- Double-click on a phonebook description or click on the Edit Phonebook... button in the Phonebook List dialog box

Use the Phonebook field at the top of the dialog box to choose a different phonebook.

The Records box contains the list of existing records in the current phonebook, in alphabetical order, either by name or company, depending on the active Sort By option. The left-hand column shows the entries in the First Name and Last Name fields; the right column shows those in the Company Name field. To add a new entry, click on the New button. The New Record dialog box shown in Figure 10.4 will appear.

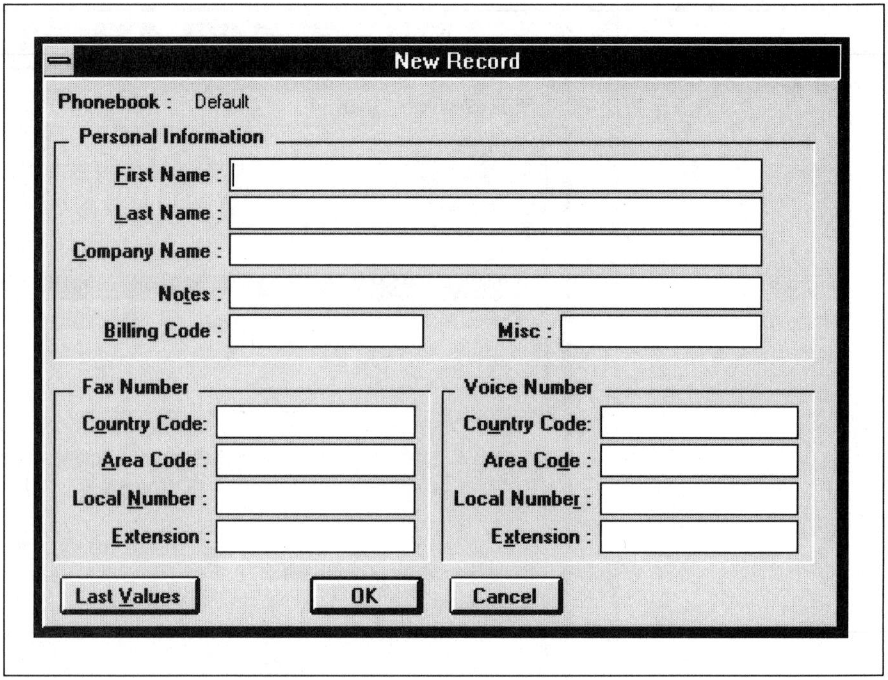

FIGURE 10.4

The New Record
dialog box

Type the name of the person or company in the appropriate fields. Since WinFax sorts names in alphabetical order by last name, you can include companies in the Name column by typing company names in the Last Name field, in addition, to typing them in the Company column. WinFax won't accept an entry unless there's information in the Last Name field.

If you type information in the wrong field, you can move it by highlighting the text you want to move with your mouse and holding down the left mouse button while you drag the information to the new field.

The Notes and Misc. fields are places for information about the current entry, such as the person's title or the reason they're in your phonebook.

(You can put any information you want into these two fields.) You can include this information when you print the listings, but it has no effect on the way WinFax handles calls to this person or company. If you use a consistent format, you can use these fields to search for records. For example, if you place the word *Client* in the Misc. field, you can search the Send Log for all faxes to clients.

The Billing Code is an account number, client code, or other identifier that some companies use to charge the costs of fax calls. When you're ready to prepare your monthly bills, you can sort the Send Log by billing code before you print it.

The Fax Number box contains the telephone number that WinFax will dial when you send a fax message to this person or company. For recipients in a different country (or, if you're in the United States or Canada, for recipients outside North America), fill in the country code and routing code. If the recipient is in a different area code from your own, type the area code. You must always fill in the Local Number field.

In some parts of North America, you must dial the area code when you place a call to another city within your own area code. If your telephone company requires an area code for these calls, place a hyphen (-) before the area code in the Area Code field. For example, if you're in Seattle and the recipient is in Tacoma, type the area code as **-206**.

If you're using a laptop or portable computer that you plan to carry with you from place to place, remember to include the area code on all entries, even if it's in your local calling area. WinFax will discard the area code when it's not needed, as long as you remember to keep the Local Area Code field up to date in the Dial Setup dialog box.

WinFax will not dial the number in the Extension field, but you might need it for reference. Since most fax machines are set up to answer automatically, you probably won't need to use this field very often.

The Voice Number fields are for information purposes only. WinFax will not dial a voice number, but it can be useful when you display or print a directory listing.

Use the Last Values button when you're making entries for several people who share the same fax number or who are in the same company. When you click on Last Values, WinFax copies the currently highlighted record. Edit the fields that are different to create a new record.

When the new entry is complete, click on OK to return to the Edit Phonebook dialog box. Click on the New... button again to create another record.

Editing an Existing Record

To make changes to an existing record, highlight the name in the Edit Phonebook dialog box and select the Edit button. The Edit Record dialog box shown in Figure 10.5 will appear. As you can see, this dialog box is identical to the New Record dialog box, but with names and numbers already filled in.

FIGURE 10.5

The Edit Record dialog box

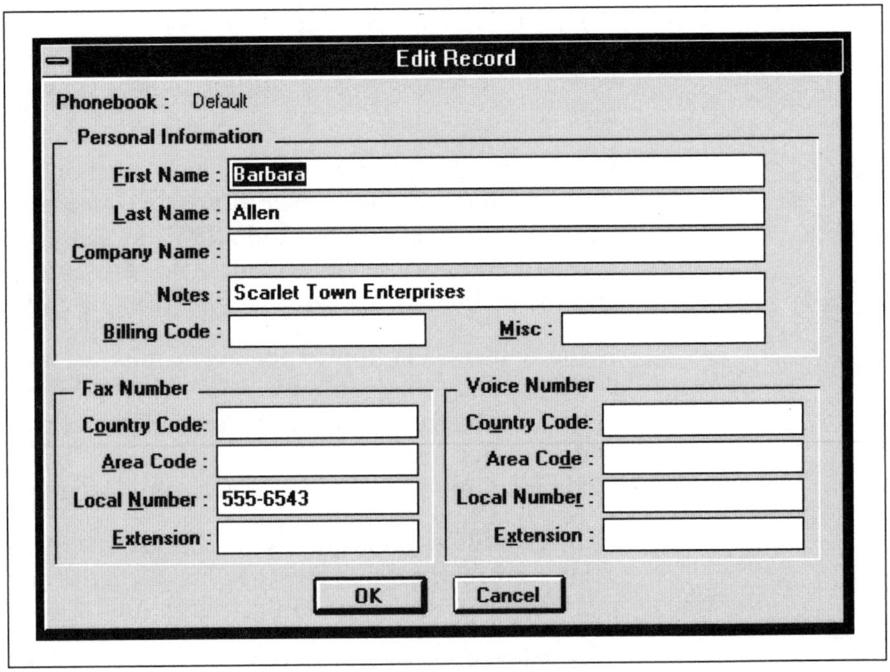

If your phonebook contains a lot of records, it might be easier to use the Search and Restore buttons to find the records you want to edit. The Search Records dialog box is identical to the Edit Record dialog box, but you only need to type into one field and then click on OK. If WinFax finds one or more matches, it will display them in an Edit Phonebook dialog box. Click on the Restore button to return to a list of all the records.

Removing a record from a phonebook is a two-step process. First, highlight the name in the Edit Phonebook dialog box and click on Delete. This will mark the record for deletion. Do the same for each record you want to remove, click on OK, and select the Phonebook List command from the Phonebook menu. Highlight the description of the phonebook that contains the records marked for deletion, click on Modify..., and finally, click on Optimize in the Modify Phonebook dialog box.

Importing a Phone List from a Database

Sometimes, you've already got a list of names and telephone numbers in a database or a text file that you want to use as a WinFax phonebook. Since WinFax stores its actual phonebook listings in the standard .DBF file format used by dBASE and many other popular database programs, you can easily import these files. It's also possible to convert from ASCII text files and files in WinFax 2.0 and CAS formats. Phonebooks imported from a database are read-only; you can't add or delete records or assign them to groups.

Importing to a New Phonebook

To import a database to a new WinFax phonebook, choose the Phonebook List command from the Phonebook menu and click on the Add button. When the Add Phonebook dialog box shown in Figure 10.6 appears, fill in the Description field as you would for a new phonebook, and either

FIGURE 10.6

The Add Phonebook dialog box

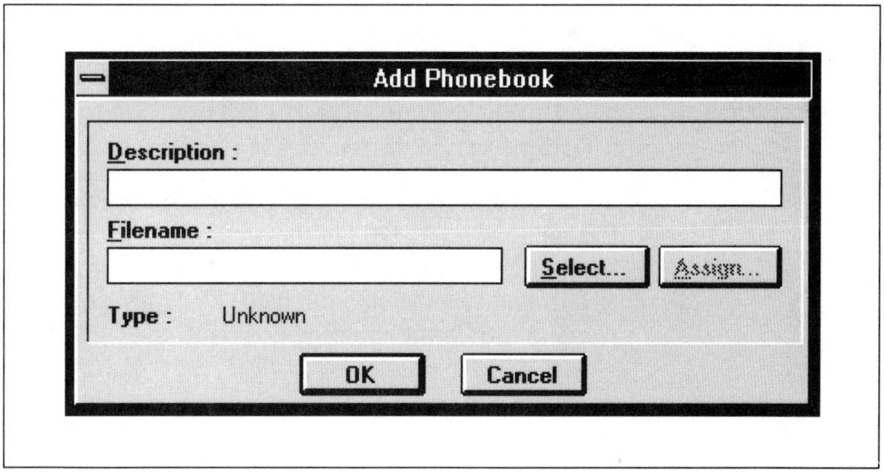

type the path and file name of the existing .DBF file you want to import or click on the Select button to use a file selector.

The Type field tells you if the file in the Filename field is a WinFax phonebook or a dBASE file. If it's a dBASE file, you must set up a cross-reference between the fields in the database and the WinFax phone-book fields. To create these cross-links, click on the Assign button to display the dialog box shown in Figure 10.7.

FIGURE 10.7

The Assign dialog box

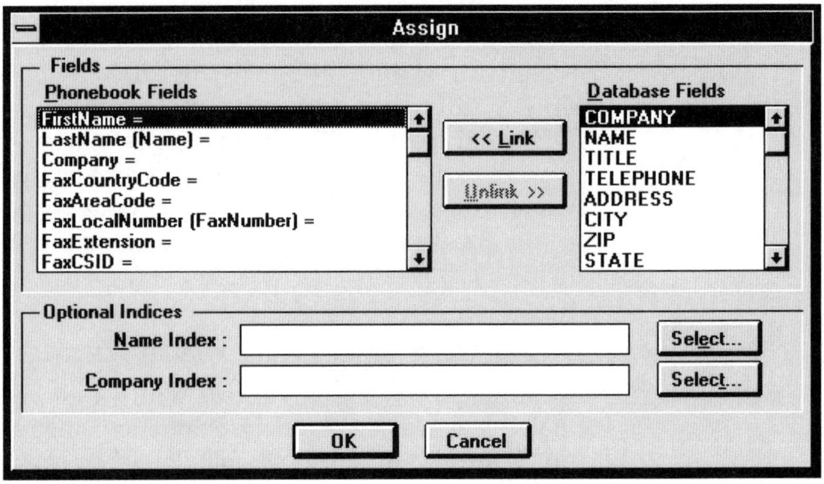

The Phonebook Fields box lists available fields in the WinFax phonebook. At an absolute minimum, you must include entries in the LastName (Name) = and FaxLocalNumber (FaxNumber) = fields.

The Database Fields box lists the fields in the .DBF file. It's not necessary to assign every database field to a phonebook field.

To set up cross-links, highlight a field in the Phonebook Fields box, highlight the corresponding field in the Database Fields box, and click on the <<Link button. If you prefer, you can click on a name in the Database Fields box and drag it to a phonebook field. The name of the linked database field appears in the Phonebook Fields box to the right of the equal sign.

To remove a cross-link, either highlight the name of the phonebook field and click on the Unlink>> button, or click on the linked field and drag it out of the Phonebook Fields box.

If you used the database program to sort records, you should import the index along with the records, since WinFax won't let you use the Sort By Name and Sort By Company options in the Edit Phonebook dialog box for imported phonebooks. Either type the path and file name of the index in the Name Index or Company Index field, or use the Select buttons to import .NDX files with a file selector. You must specify a Name Index to use partial matches.

After you complete the Assign dialog box, click on OK to return to the Add Phonebook dialog box. Click on that OK button to add the new phonebook to your phonebook list.

To change the description of an existing phonebook, highlight the old description in the Phonebook List dialog box and select the Modify button.

Importing Records into an Existing Phonebook

To import records from a text file or database file, select Import from the Phonebook Menu. The Import Phone Records dialog box shown in Figure 10.8 will appear.

The Phonebook field identifies the currently active phonebook, which is where WinFax will place imported records. To choose a different phonebook, click on the arrow to open a file selector. To place the imported records in a new phonebook, select the Create button and type the path and file name of the new phonebook in the New Phonebook dialog box.

The Import From field identifies the source file that contains the records. Either type the path and file name in the field, or click on the Select button to use a file selector.

The Format field shows the format of the source file. Click on the arrow to select a different format. If you specify CAS or WinFax 2.0, the Options box will disappear. If you specify dBASE, the lower half of the dialog box will change.

If the source file is in ASCII format, it will be easier to specify the format if you print a couple of sample records. In order to import records from an ASCII file, you must identify the order in which the fields appear in

FIGURE 10.8

The Import Phone
Records dialog box

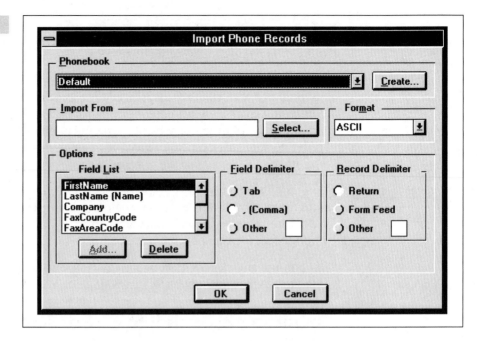

the ASCII file and the delimiters used to separate fields within a record
and to separate one record from the next.

To change the order of the Field List, click on a field name and drag the
field to its correct position. To delete a field name, highlight it in the Field
List and click on the "Delete" button.

If the source file is a dBASE file, the dialog box will look like the one
shown in Figure 10.9. The Phonebook Fields box lists available fields in
the WinFax phonebook. At an absolute minimum, you must include entries in the LastName (Name) = and FaxLocalNumber (FaxNumber) =
fields.

The Import Fields box lists the fields in the dBASE file. It's not necessary
to assign every database field to a phonebook field. To select a field in the Import Fields list, highlight the field name, highlight the name in the Phonebook Fields box that you want to assign to that field, and click on the <<Link
button. If you prefer, you can click on a name in the Import Fields box and
drag it to a phonebook field. The name of the linked database field appears
in the Phonebook Fields box to the right of the equal sign.

FIGURE 10.9

The Import Phone
Records dialog box for
a dBASE file

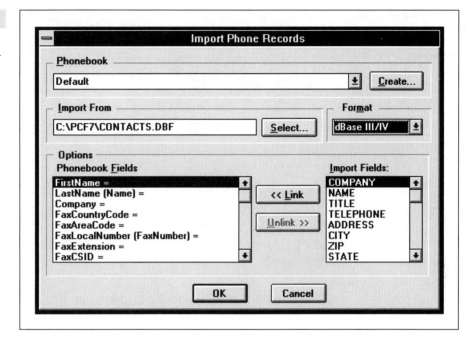

To remove a cross-link, either highlight the name of the phonebook field
and click on the Unlink>> button, or click on the linked field and drag it
out of the Phonebook Fields box.

Making Other Changes to WinFax 3.0 Phonebooks

In addition to adding and deleting records, you can also change the description that appears in the Phonebook List and define a different
phonebook as the default.

Changing a Phonebook Description

To change a description, select the Modify button in the Phonebook List
dialog box and type the new description in the Modify Phonebook dialog
box shown in Figure 10.10.

FIGURE 10.10

The Modify
Phonebook dialog box

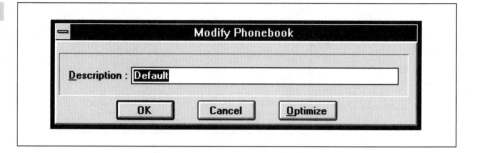

Changing the Default Phonebook

To change the default phonebook, open the Phonebook List dialog box. Highlight the description of your new default phonebook, and click on the Set Default button. The telephone icon will move to the new default phonebook.

Removing a Phonebook from Your List

To remove a phonebook from the list, highlight the description of that phonebook, and click on the Delete button. If the highlighted phonebook is linked to a database, the Delete button changes to a Remove button.

Creating Phonebook Groups

When you create a group, you can enter a single command to send the same fax to every member of the group. If you use variables in the cover-sheet, you can even personalize the cover sheet for each recipient.

To create a new group, either click the PhnBook icon button or select the Records... command in the Phonebook menu to open the Edit Phonebook dialog box , and click on the Groups button. When the Groups dialog box shown in Figure 10.11 appears, select the New... button.

The New Group dialog box shown in Figure 10.12 has only one field. Type the name of the new group and click on OK.

To add records to a group, highlight the names in the Records list and the name of the group in the Groups list. Then click on the Add>> button. The names will appear under the name of the group. You can place the same record in more than one group.

FIGURE 10.11

The Groups dialog box

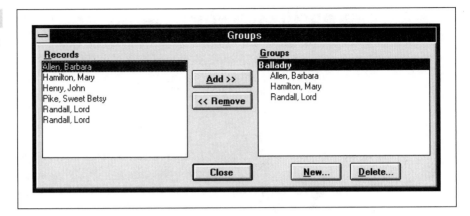

FIGURE 10.12

The New Group
dialog box

If you want to select a series of adjacent records, hold down the Shift key and click on the first and last names that you want to select. To select several unrelated records at once, hold down the Ctrl key and click on each record. Then drag all of the highlighted records to the name of a group in the Groups list.

To remove records from a group, highlight the records in the Groups list and click on the <<Remove button.

To remove a group, highlight the group name and click on the Delete… button.

When you're finished assigning records to groups, click on Close.

Exporting a Phonebook to a Text File

You can print your phonebook directly from WinFax, but in some cases it may be more convenient to work with phonebook records in your word processor, database, or other applications. The easiest way to do this is to export a phonebook to an ASCII text file.

> **NOTE**
>
> WinFax Pro and WinFax Lite don't use the same file format for phonebooks. If you're using WinFax Pro on one computer and WinFax Lite on a second machine, you must convert the WinFax Pro phonebook to WinFax 2.0 format.

To create an ASCII text file or a WinFax 2.0 phonebook, select the Export command from the Phonebook menu to display the Export Phone Records dialog box shown in Figure 10.13.

When the dialog box appears, the current phonebook description is in the Phonebook field. If you want to export a different phonebook, open the list and select a different description.

FIGURE 10.13

The Export Phone Records dialog box

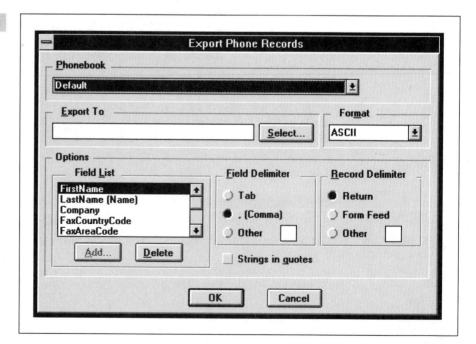

Use the Export To field to specify the destination file. Either type the path and file name, or use the Select button to choose an existing file.

In the Format field, you can choose either ASCII or WinFax 2.0 (which is the same as WinFax Lite). If you choose the WinFax format, the Options box will disappear.

If you're creating an ASCII file, the Options box specifies the content and format of each record. If you don't want to include a field from the WinFax phonebook in the ASCII file, highlight the name of that field in the Field List and click on the Delete button. If you change your mind, click on the Add button and choose the field name from the list of deleted fields.

To change the order of fields, either drag the name of the field to its new location with your mouse; or delete the field name, highlight the new location, and use the Add button to replace it.

If you're exporting your phonebook list to a database, you must specify the delimiters that separate fields within each record and that keep records separate. Check your database documentation for details about field and record delimiters. If your database requires quotation marks around text strings, select the Strings In Quotes option.

When all the information in the dialog box is accurate, click on OK to create the new file.

Printing Information from a Phonebook

Sometimes it's hard for a committed computer user to admit it, but there are times when the best way to store or distribute information is on paper. WinFax allows you to organize the information in a phonebook for printing in a variety of ways. You can print directly from WinFax, and you can export a phonebook to your word processor or desktop publisher for more complex formatting or to incorporate the phonebook into a larger document.

To print a phonebook from WinFax, select the Print command from the Phonebook menu. The Print Phonebook menu shown in Figure 10.14 will appear.

The Printer field identifies the printer where WinFax will print the phonebook. To select a different printer (including WinFax), click on the Setup... button.

FIGURE 10.14

The Print Phonebook menu

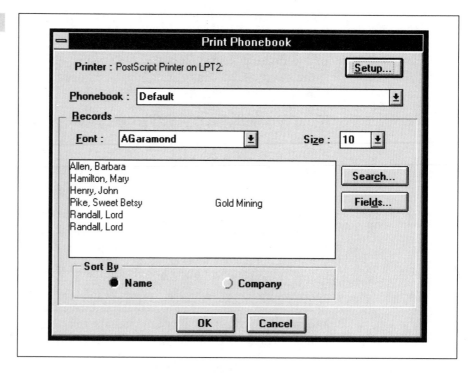

The description of the current phonebook appears in the Phonebook field. To select a different phonebook, click on the ↓ button to the right of the description.

The Font and Size fields specify the font and type size your printer will use to print the phonebook. All text will print in the same font and size. If you need to use multiple fonts or sizes, export the phonebook to a word processor or desktop publisher and use that application's formatting and printing tools.

To print every record in the phonebook, select either Sort By Name or Sort By Company to specify the order in which the records will print. To print selected records, click on the Search button to display the Search Records dialog box shown in Figure 10.15.

Fill in the fields that define the records you want to print. For example, since you cleverly identified records in the Misc. field as Customer or Supplier, you can tab to that field in the Search Records dialog box and type **Customer** to create a complete list of Customers, while ignoring all the other records in the phonebook.

FIGURE 10.15

The Search Records
dialog box

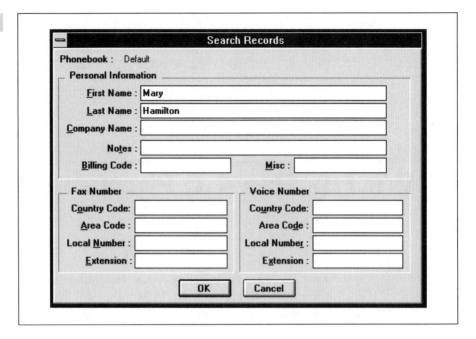

You probably don't need to print every field in your phonebook records.
Click on the Fields button to display the Print Fields dialog box shown
in Figure 10.16.

FIGURE 10.16

The Print Fields
dialog box

Select the fields that you want to include in your printout and use the ↑ and ↓ buttons to specify the number of characters in each field. One Line Per Record specifies whether to print the whole record on a single line or to place each field on a new line. Click on OK to accept the list of fields.

When all the information in the Print Phonebook dialog box is complete, click on OK to start printing.

WinFax Pro 4.0 Phonebooks

Phonebooks in WinFax Pro 4.0 are similar to the ones in earlier versions of WinFax, but since they use the new Multiple Document Interface, the screens you see on your computer are very different.

The Phonebooks window is the starting point for creating, modifying, and removing phonebooks in WinFax 4.0. Either click on the Phnbooks button or select Phonebooks... from the Fax menu to start working on phonebooks. The Phonebooks window shown in Figure 10.17 will appear within the Delrina WinFax PRO screen.

The folder list on the left side of the Phonebooks screen contains a list of existing phonebooks and groups. You can open a phonebook by clicking on its name. Select New Phonebook/Group to add a phonebook to the list or to create a new phonebook from scratch. If you keep all your names and fax numbers in a single phonebook, you can hide the phonebook list by by dragging the edge of the current phonebook box to the left.

When you click on New Phonebook/Group, the Add Phonebook/Group dialog box shown in Figure 10.18 appears. If you plan to enter each entry in the phonebook by hand, choose the Add New dBASE option; to import a phonebook from an existing database, select the Use Existing Add File option.

In the Name field, type the name that you want WinFax to use in the list of phonebooks. Unlike most computer lists, you can enter descriptions in plain language (up to 40 characters). In the Type box, select Phonebook,

FIGURE 10.17

The WinFax Pro 4.0
Phonebooks screen

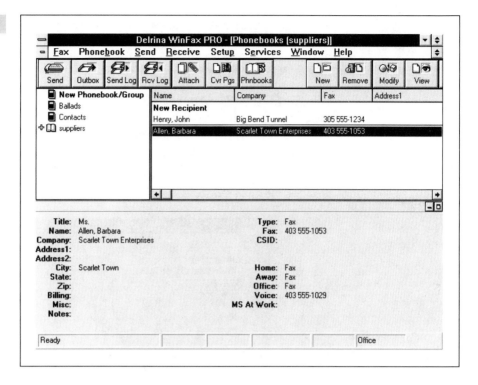

FIGURE 10.18

Add
Phonebook/Group
dialog box

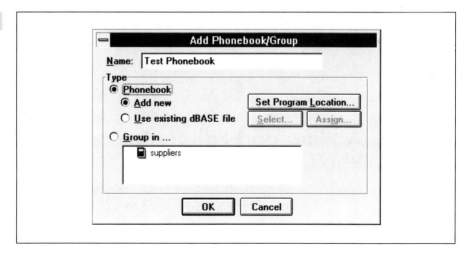

and choose either Add New to create a new phonebook, or Use Existing dBASE File to import the phonebook from a data base.

Click on the Set Program Location... button (shown in Figure 10.19) to specify the names and locations of the DOS files that WinFax will use internally for this phonebook. The default File Name Prefix is the first seven letters of the phonebook's name, and since you'll never work with those internal files, you should accept the default, unless you're already using that default for another phonebook.

FIGURE 10.19

The Set Program Location dialog box

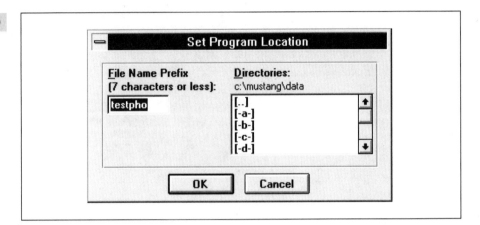

The default directory for phonebook files is the DATA subdirectory in the directory that contains the WinFax files. In most cases, that's a good place for them, but if you want to store them someplace else, such as on a floppy disk or a network drive, use the Directories field to specify the new path.

Adding and Editing Records in a Phonebook

When you create a new phonebook, it does not contain any names or fax numbers. There are four ways to add entries to the current phonebook:

- Highlight New Recipient at the top of the list of names in the Phonebooks window. If the phonebook is empty, New Recipient will be the only item in the item list.

- Highlight any item in the item list, and then click on the New button in the button bar.

- Highlight any item in the item list, and select the New... command from the Phonebooks menu.

- Select a phonebook description from the phonebook list and click with your right mouse button to display a menu. Select the New command from that menu.

When you use one of these techniques to add an entry to a phonebook, the New Recepient dialog box in Figure 10.20 will appear.

You can make changes to an existing phonebook record by either double-clicking on the entry in the phonebook or by highlighting the entry and clicking on the Modify icon button. The Modify Record dialog box is identical to the New Recepient dialog box.

The title bars of the New Recepient and Modify Record dialog boxes identify the phonebook where they will store the current record. If you type information in the wrong field, you can move it by highlighting the text with your mouse and then holding down the left mouse button while

FIGURE 10.20

The New Recipient
dialog box

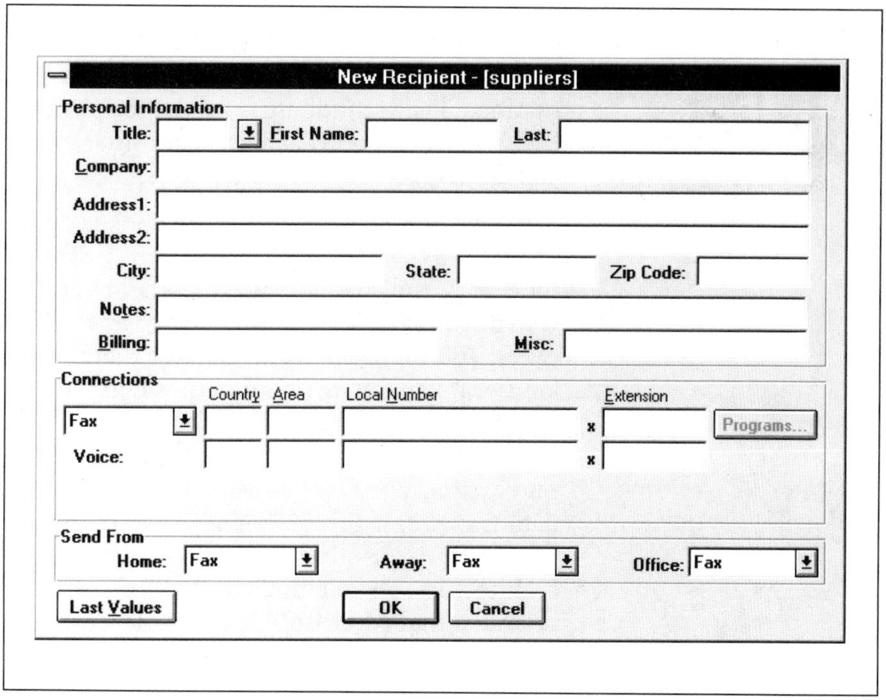

you drag the information to the new field. The New Recepient and Modify Record dialog boxes include these fields:

Personal Information The First Name and Last fields are for the name of the person who will receive faxes when you select this phonebook entry. Use the Company field to identify the name of the business. You will be able to organize your phonebook by either last name or company name by using the Sort By commands in the Phonebooks menu.

The Notes and Misc. fields are places for additional information about the current entry. It really doesn't matter what you put into these two fields. You can include this information when you print the listings, but it has no effect on the way WinFax handles calls to this person or company. If you use a consistent format for information you enter in these fields, you can use these fields to search for records. For example, if you place the word *Client* in the Misc field, you can search the Send Log for all faxes to clients.

The Billing code is an account number, client code, or other identifier that some companies use to charge the costs of fax calls. When you're ready to prepare your monthly bills, you can sort the Send Log by billing code before you print it.

Connections The Telephone box contains space for two telephone numbers, including country code, area code, and extension number. The second line is for the recipient's voice telephone number. WinFax won't dial a voice number, but it can be useful when you display or print a directory listing.

The first line of the Connections box is for the number that Win-Fax will dial when you send a fax to this person or company. At the left side of this line, there's a field with a drop-down menu. Use this field to specify the type of link you expect to establish between your computer and this recipient. If you don't know what kind of fax machine or modem is connected to this telephone number, select Fax. If you know that this recipient is using a modem configured for binary file transfer (BFT) or has another computer using WinFax Pro Version 4.0, choose that option for faster data transfer. If this recipient is a computer, fax machine, fax server, network, or other device that supports the Microsoft At Work architecture, choose the MS At Work option. Microsoft has announced that Microsoft At Work desktop software will become a standard part of future versions of the Windows operating system, so this alternative is likely to become a lot more common in the next few years. Chapter 16 contains more information about using WinFax to transmit binary files.

When you use the BFT, WinFaxBFT, or MS At Work options, you can transfer files through your fax modem without coverting them to a fax image, if both your computer and the system that receives the file contain applications that can recognize the files. Since WinFax can transfer binary files more quickly than it can transfer images (because it doesn't have to convert it to a bitmap image), you will want to use one of the binary transfer methods whenever possible. If your recipient will be incorporating the material you send into another document, the file transfer technique is much more efficient (and probably more accurate) than optical character recognition.

Some form of binary file transfer will probably eventually become standard, but right now, you're probably going to have to telephone the person to whom you want to send the files and spend some time coordinating your computer with the recipient's system. You'll need to find out which form of fax file transfer the recipient can support and which Windows applications files they can recognize. If the person to whom you're sending the file doesn't support any kind of fax file transfer, you may want to use your data communications program to send the file, rather than using WinFax at all.

Use the Programs... Programs button to display the Available At Recipient dialog box shown in Figure 10.21, which contains a list of the Windows applications on your computer in the Possible Programs box. Highlight the applications that are also on the recipient's system and click on the Add>> button. Remember that many Windows word processing, spreadsheet, and database applications have filters that can recognize files created with competing programs; for example, if you use Word for Windows as your word processor and your recipient uses Ami Pro, you should place the winword.exe file in the Available list. When you have placed all the shared applications in the Available field, click on OK.

When you select the MS At Work option, an Alias field appears in the Telephone box. If the recipient has a Microsoft At Work alias, type it in this field.

FIGURE 10.21

Programs Available
dialog box

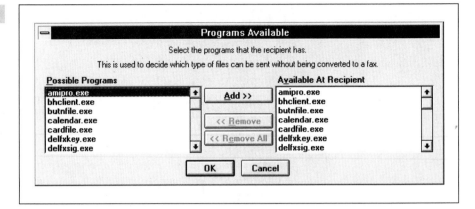

Send forms   The Default fields specify the type of network transmission that WinFax will use to send a fax or file to this recipient. You can configure the same phonebook entry to use different networks when the recipient is at home, in her office, or traveling.

If your phonebook contains a lot of records, it might be easier to use the Search and Restore commands in the Phonebooks menu to find a record to use either to send a fax or file or to edit. The Search dialog box is identical to the Modify Record dialog box, but you only need to type one or more complete or partial fields and click on OK or press ↵. WinFax will display the phonebook entries that match your specification. To return to a display of all items in the current phonebook, use the Restore command in the Phonebooks menu.

To delete an item from a phonebook, highlight the entry and click on the Remove icon button, or use the Remove command in the Phonebooks menu. You can also display the Phonebooks menu by clicking on your right mouse button when the cursor is on the highlighted phonebook record and selecting the Remove command.

Importing a Phone List from a Database

If you have a database that includes names and fax numbers, you can use that list as a WinFax phonebook. Since WinFax stores its phonebook listings in the standard .DBF format used by dBASE and many other popular database programs, you can import these files to a Phonebook. You can

also convert from ASCII text files and files in older WinFax and CAS formats. Phonebooks imported from a database are read-only; you can't add or delete records or assign them to groups.

Importing to a Phonebook

There are three ways to import a database to WinFax:

- Select the New command from the Phonebooks menu
- Double-click on New Phonebook/Group in the phonebook list on the left side of the screen
- Highlight a phonebook, click with your right mouse button to display the Phonebooks menu, and select the New command.

When the Add Phonebook/Group dialog box appears, type a name for this phonebook in the Name field, and select the Use Existing dBASE File for phonebooks option and a click on Select.... The Select Phonebook/Database File dialog box in Figure 10.22 will appear.

FIGURE 10.22

The Select Phonebook/Database File dialog box

```
Select Phonebook/Database File

File Name:                    Directories:              OK
[*.dbf]                       c:\mustang\data
                                                        Cancel
ballad.dbf                    c:\
balladg.dbf                     mustang
default.dbf                     data
defaultg.dbf
presrel.dbf
presrelg.dbf
suppli1.dbf
suppli1g.dbf
supplie.dbf                   Drives:
supplieg.dbf                  c: stacvol_dsk

List Files of Type:
Delrina WinFax and dBASE III/IV (*.dbf)
```

Either type the path and file name in the File Name field, or choose the database file from the list.

The List Flies of Type field tells you if the file in the Filename field is either a WinFax phonebook or a dBASE file. If it's a dBASE file, you must set up a cross-reference between the fields in the database and the WinFax phonebook fields. To create these cross-links, click on the Assign button to display the dialog box shown in Figure 10.23.

FIGURE 10.23

The Data Field Assignment dialog box

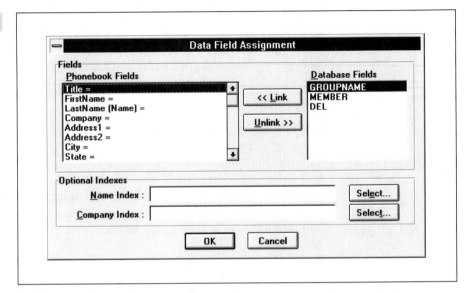

The Phonebook Fields box lists available fields in the WinFax phonebook. At an absolute minimum, you must include entries in the LastName (Name) = and FaxLocalNumber (FaxNumber) = fields.

The Database Fields box lists the fields in the .DBF file. It's not necessary to assign every database field to a phonebook field.

To set up cross-links, highlight a field in the Phonebook Fields box, highlight the corresponding field in the Database Fields box, and click on the <<Link button. If you prefer, you can click on a name in the Database Fields box and drag it to a phonebook field. The name of the linked database field appears in the Phonebook Fields box to the right of the equal sign.

To remove a cross-link, either highlight the name of the phonebook field and click on the Unlink>> button, or click on the linked field and drag it out of the Phonebook Fields box.

If you used the database program to sort records, you should import the index along with the records, since WinFax won't let you use the Sort By Name and Sort By Company commands in the Phonebooks menu when you're using an imported phonebook. Either type the path and file name of the index in the Name Index or Company Index field, or use the Select button to choose .NDX files from a file selector.

After you complete the Assign dialog box, click on OK to return to the Use Phonebook File dialog box. Make sure there's an entry in the Description field, and click on that OK button to add the imported phonebook to your phonebook list.

Importing Records from Text Files and Old WinFax Phonebooks

To import records from a text file or a WinFax Pro 3 or WinFax Lite phonebook, highlight the name of the phonebook where you want the records to go and select the Import command from the Fax menu. The Import Phone Records dialog box in Figure 10.24 will appear.

The Import From field identifies the source file that contains the records. Either type the path and file name in the field, or click on the Select... button to use a file selector.

FIGURE 10.24

The Import Phone Records dialog box for ASCII

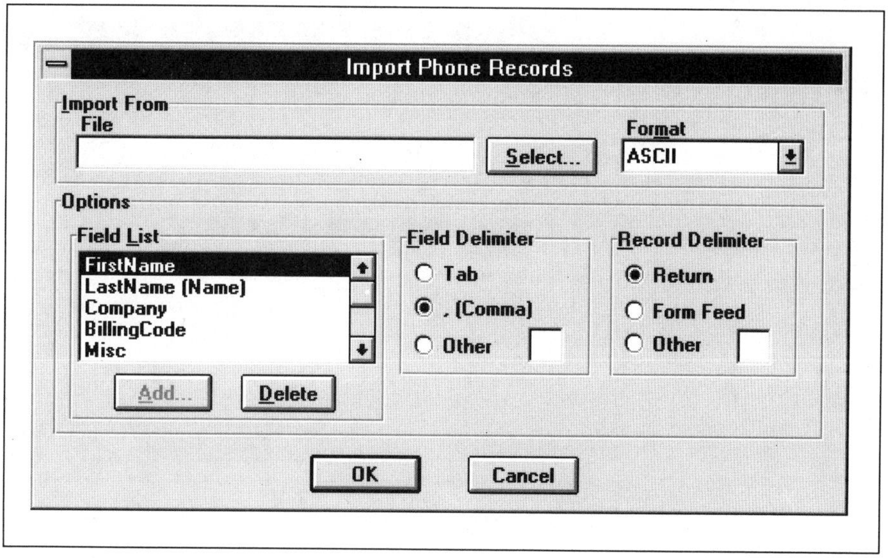

The Format field shows the format of the source file. Drop down the menu to select a different format. If you specify CAS or WinFax 2.0, the Options box will disappear. If you specify dBASE III/IV, the lower half of the dialog box will change.

If the source file is in ASCII format, use Windows Write or some other text editor to print a couple of sample records before you start to work on the Import Phone Records options. In order to import records from an ASCII file, you must identify the order in which the fields appear in the ASCII file and the delimiters used to separate fields within a record and to separate one record from the next.

To change the order of the Field List, click on a field name and drag the field to its correct position. To delete a field name, highlight it in the Field List and click on the Delete button.

If the source file is in an older WinFax format, the Import Phone Records dialog box will not include an Options field.

If the source file is a dBASE file, the dialog box will look like the one in Figure 10.25. The Phonebook Fields box lists available fields in the Win-Fax phonebook. At an absolute minimum, you must include entries in the LastName (Name) = and FaxLocalNumber (FaxNumber) = fields.

FIGURE 10.25

The Import Phone Records dialog box for dBASE files

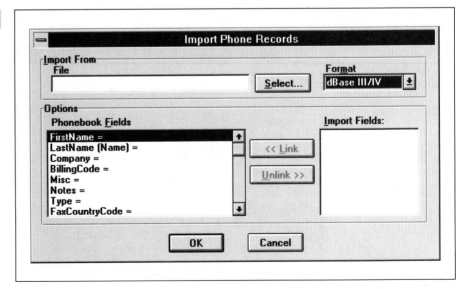

The Import Fields box lists the fields in the dBASE file. It's not necessary to assign every database field to a phonebook field. To select a field in the Import Fields list, highlight the field name, highlight the name in the Phonebook Fields box that you want to assign to that field, and click on the <<Link button. If you prefer, you can click on a name in the Import Fields box, and drag it to a phonebook field. The name of the linked database field appears in the Phonebook Fields to the right of the equal sign.

To remove a cross-link, either highlight the name of the phonebook field and click on the Unlink>> button, or click on the linked field and drag it out of the Phonebook Fields box.

Removing and Renaming Phonebooks in WinFax Pro 4.0

To change the name of a phonebook click on the Modify icon button, or use the Modify command in the Phonebooks menu. When the Modify Phonebook dialog box shown in Figure 10.26 appears, edit the name of the phonebook in the Name field.

FIGURE 10.26

The 4.0 Modify Phonebook dialog box

Modify Phonebook
Name: Ballads
Type
● **P**honebook
○ **A**dd new   Set Program Location...
● **U**se existing dBASE file   Select...   Assign...
○ **G**roup in ...
OK   Cancel

To mark an entry to remove it from a Phonebook list, first highlight that entry, then click on the Remove icon button or select Remove from the Phonebook menu. After you have marked all the entries you want to remove, select the Optimize command from the Phonebook menu.

To remove a phonebook from the phonebook list, highlight the name of the phonebook and either click on the Remove icon button, or use the Remove command in the Phonebooks menu or the right mouse button menu.

Creating Phonebook Groups

Groups are lists of people to whom you want to send the same fax. If you place variables in the cover sheet, you can personalize the cover sheet for each recipient.

To create a new group, select New Phonebook/Group from the list of phonebooks on the left side of the Phonebooks window. When the Add Phonebook/Group dialog box appears, type the name of the group in the Name field, click on Group In..., and select the phonebook that should contain the group.

The list of phonebooks now includes the new group as a sub-entry under the phonebook that contains it. You can hide it by highlighting the name of the phonebook (not the group name), and selecting Collapse Folder in the Phonebooks Menu.

To see groups in a collapsed folder, select Expand folder in the Phonebooks menu.

Exporting a WinFax 4.0 Phonebook to a Text File

The procedure for exporting WinFax 4.0 phonebooks to a text file and for printing information from a phonebook is exactly the same as WinFax 3.0. See "Exporting a Phonebook to Text File" and Printing Information from a Phonebook" above for details.

Once you have created a phonebook record, you can send a fax message to the name and telephone number in that record by typing the first few letter s of the name in the Send Fax dailog box. We'll explain the process of sending faxes in the next chapter.

CHAPTER
11

Sending Fax Messages with WinFax Pro

THERE are two ways to send faxes with WinFax: You can fax documents directly from Windows applications the same way you send them to your printer, or you can use the WinFax Administrator to select and send fax files that you created earlier.

In most cases, it's easier to enter a print command from your word processor, spreadsheet, graphics program, or other application.

This chapter contains complete instructions for both techniques. We'll also explain how to add cover sheets, combine documents from different applications into a single fax message, and schedule a delayed fax transmission.

Sending Faxes from Windows Applications

As far as your Windows applications are concerned, WinFax is just one more printer. After you create a document (which could contain text, graphics, or any combination of the two), all you need to do to send a fax is to enter the Print or Printer Setup command and select WinFax from the list of printers.

In most applications, the print commands are in the drop-down File menu. When you "print" one or more pages to WinFax, the fax program converts each page to a separate image file, which you can send immediately, schedule for delayed transmission, or store as a file without sending it. WinFax 4.0 can also transfer files to another computer without converting them to images. If both systems are able to recognize the same application data files, this new feature can improve transmission speed and reduce the amount of disk space required to send faxes.

If you expect to send faxes more often than you create paper copies on your own printer, you should consider making WinFax your default printer. But if you share your computer with other users, this could be a fine recipe for confusion. You can change the default printer by opening the Windows Print Manager and selecting Options ➤ Printer Setup... When the Printers dialog box shown in Figure 11.1 appears, highlight WINFAX On COM# and click on the Set As Default Printer button.

FIGURE 11.1

The Windows Print Manager Printers dialog box

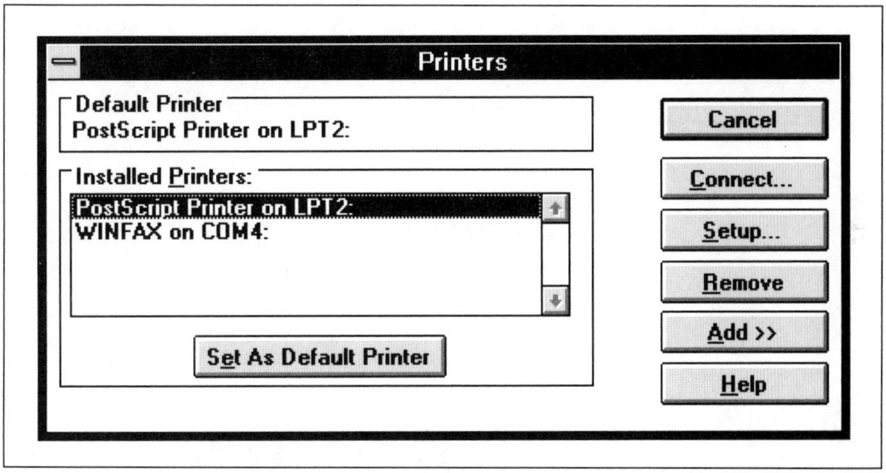

If you print to paper more often than you send faxes, your printer should be the default. Either way, when you want to send a fax from an application or Windows utility, make sure WinFax is the current printer.

For example, Figure 11.2 shows the File menu in Windows Write. When you select Print Setup, you will see the dialog box shown in Figure 11.3.

If WINFAX is not your default printer, click on Specific Printer and select WINFAX On COM# from the file selector. The exact procedure is slightly different in different Windows applications and programs, but in general, look for a Print Setup option in the File menu or Options menu, or a Setup button in the application's Print dialog box. If you're working in one of those odd applications that forces you to use the default printer, such as some versions of FrameMaker, use the Windows Print Manager to change the default before you try to send a fax.

FIGURE 11.2

The File menu in
Windows Write

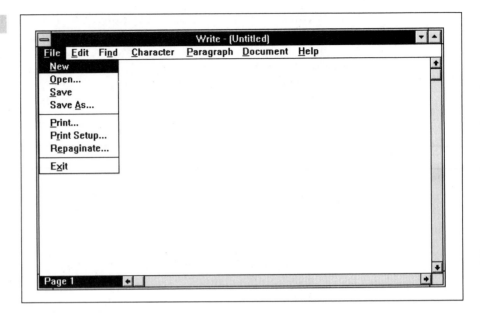

FIGURE 11.3

The Windows Write
Print Setup dialog box

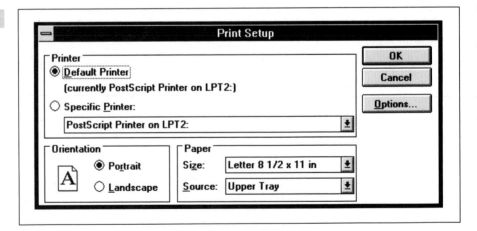

When WinFax is your current printer, use the application's formatting
tools to define the layout of each page. Don't worry about a cover sheet
right now; WinFax will add one later if you wish. When you're satisfied
with the appearance of the document, choose Print from the application's
File menu. If the application displays a Print Options dialog box, choose

the options you want to use for this document and click on OK or the Print button. Instead of sending the document directly to a printer, the Windows Print Manager passes the document over to WinFax, which displays the Send Fax dialog box. We'll explain how to fill in the Send Fax dialog box later in this chapter.

After you complete the Send Fax dialog box and click on the Send button, Windows will return to your original application, and WinFax will transmit your fax in background. If Display Call Progress is active in the Win-Fax Program Setup dialog box, a status message will appear on your screen during transmission.

At this point, you can continue working in the application from which you sent the fax, or close the application and do something else. However, Windows and the Print Manager must continue to run until the transmission is complete, so you can't exit Windows or turn off the computer.

When you send a fax through WinFax from a Windows application, Win-Fax creates and stores a fax image file for each page. If you know that you will want to combine pages from more than one application into a single fax message, you can treat the documents from each application as attachment files. We'll explain that process later in this chapter.

Sending Faxes Directly from WinFax

It's also possible to send faxes directly from WinFax, but in WinFax Pro 3.0, you can only send files that are already in WinFax attachment file format. In practice, this means the only things you can send directly from WinFax are documents that you've previously converted to Winfax format or sent to the same or another recipient, and one-page notes on cover sheets. Since WinFax saves each page of a multi-page document as a separate file, you can select the pages that you send, rather than retransmitting the whole thing.

Here's the procedure for sending directly from WinFax:

1. If it's already running, switch to WinFax. If it's not running, start the program.

2. Either click on the Send icon or select Send ➤ Fax…. One of the Send Fax dialog boxes shown in Figure 11.4 and 11.5 will appear.

3. Fill in the Send Fax dialog box. The next two sections of this chapter explain the Send Fax dialog box in detail. Click on the Add button in the Attachments box in WinFax Pro 3.0, or the Attach button in WinFax Pro 4.0 to select the files that you want to send. Use options in the Cover Page box to choose a cover sheet and include a message on the cover sheet.

FIGURE 11.4

The WinFax Pro 3.0 Send Fax dialog box

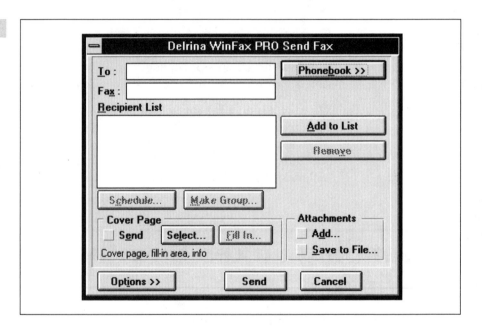

4. Click on the Send button to send the attachment files, with or without a cover sheet, or to send a cover sheet with no attachments.

The Send Fax Dialog Box

As the name suggests, the Send Fax dialog box controls the way WinFax sends faxes. It includes the name and fax telephone number of each recipient, and specifies whether the transmitted fax message will include a

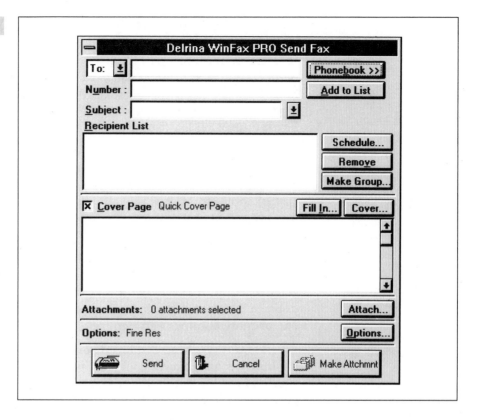

cover sheet or attachments. You can also change the default settings for dialing prefix and resolution, and perform other special functions from the Send Fax dialog box.

Specifying the Recipient

The first part of the Send Fax dialog box specifies the name and fax number of the person or company to whom you want to send a fax message. If you wish, you can send the same fax to more than one recipient, either by choosing a phonebook group or by creating a recipient list.

The Send Fax dialog box includes these fields:

To   This field contains the name of the person or company to whom you want to send the fax. You can either type the name as "First Name<space>Last Name" or as "Last Name,<space>First

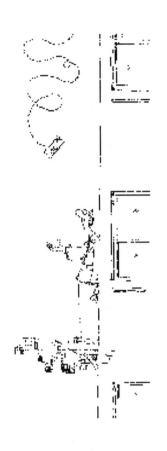

Name." If the recipient is a company rather than an individual, type the company name.

If Use Partial Matches In Phonebook is active in the Program Setup dialog box, WinFax will search through the current phonebook for a last name that matches the first two or three letters you type, and will automatically fill in the rest of the name and the Fax Number field.

In WinFax Pro 4.0, you can send a duplicate copy of the same fax to another recipient by opening the menu at the To field name and selecting cc: (for *carbon copy*) instead. You can choose recipients of copies from your phonebook the same way you specify the main recipient.

Fax (WinFax Pro 3.0) or **Number** (WinFax Pro 4.0)   The Fax number is the telephone number of the recipient's fax machine or modem. WinFax will dial the telephone number in this field, exactly as it appears here (ignoring spaces and dashes). If you need to pause between digits in order to wait for a second dial tone, use one or more commas between those digits.

When you type a long-distance fax number, remember to include the long-distance access code and area code. For example, if you're in the United States or Canada, you should place a 1 ahead of the area code for calls to North America, or 011 ahead of an overseas telephone number. Don't worry about the access code when you're adding a number to the phonebook—WinFax takes care of this for you.

If you're calling from an office where you need to dial 9 for an outside line on every call or some other prefix for a long-distance line, click on the Options button and fill in the Use Prefix field in the Options section or the Send Options dialog box. Many businesses have telephone systems that can choose less expensive long-distance routing if you dial a special prefix. If you're not sure about this, ask the person responsible for maintaining your telephones.

Phonebook>>   To select one or more names and fax numbers from a phonebook or to add the current name to a phonebook, click on Phonebook>>.

The Phonebook>> button instructs WinFax to expand the Send Fax dialog box to include information about your phonebook, as shown in Figures 11.6 and 11.7. Figure 11.6 shows the expanded Send Fax dialog box in WinFax Pro 3.0. Figure 11.7 shows the same dialog box in Win Fax Pro 4.0. The next time you click on the Phonebook>> button, the expanded part of the dialog box will disappear.

FIGURE 11.6

The WinFax Pro 3.0 Send Fax dialog box with the Phonebook section visible

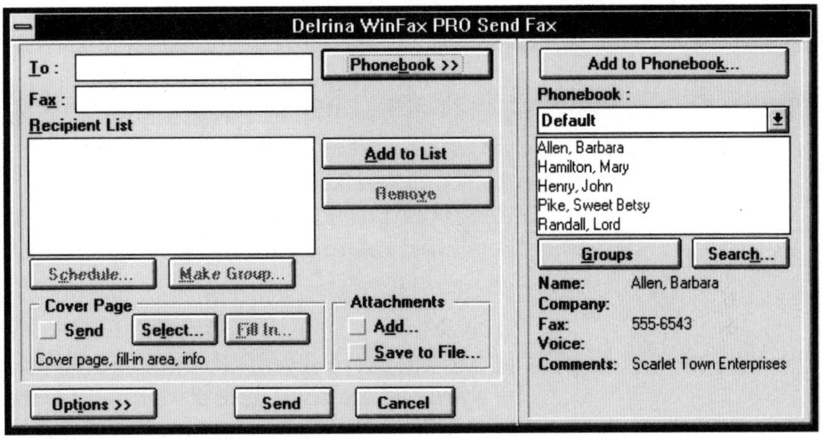

FIGURE 11.7

The WinFax Pro 4.0 Send Fax dialog box with the Phonebook section visible.

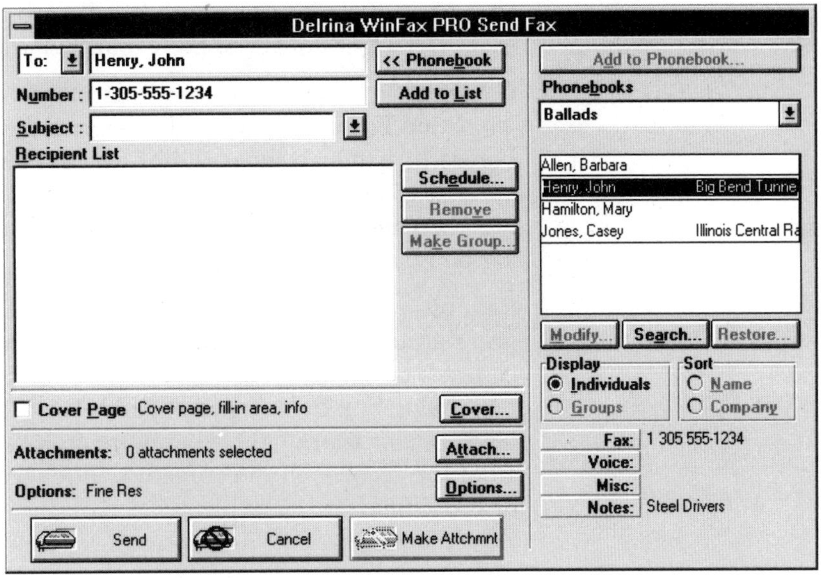

Filling in the WinFax 3.0 Expanded Phonebook Section

The section of the Send Fax dialog box that appears when you click on the Phonebook>> button in Version 3.0 is quite different from the same dialog box in Version 4.0. If you're using WinFax Pro 3.0, follow the instructions in this section; if you're using WinFax Pro 4.0, skip to the next section of this chapter.

> **Add to Phonebook**...   If you've already typed a name and fax number in the To and Fax fields, you can add that name to the current phonebook by clicking on the Add To Phonebook... button. When you select Add To Phonebook..., the New Record or Add Record dialog box appears, with the name and fax number fields filled in. You can find a complete explanation of this dialog box in Chapter 10.
>
> **Phonebook**   The Phonebook field shows the description of the current phonebook. To choose another phonebook, click on the arrow at the right side of the box to drop a selector list.
>
> The names in the box under the phonebook description are from the current phonebook. If the phonebook has more than five names, use the up or down arrows to display additional names. When you highlight a name to select it, detailed information about that recipient will appear in the To and Fax fields, and in the lower-right corner of the expanded dialog box.
>
> **Groups**   Click on Groups to display a list of phonebook groups. The Send Fax dialog box will change to the arrangement shown in Figure 11.8, which shows a list of groups and the members of the current group. To send the same fax to all members of a group, highlight the group name. Click on Individuals to change the dialog box back to its original format.
>
> **Search**   If you have a large phonebook, it can be tedious to find names by scrolling through the list one name at a time. You can reduce the number of names that actually appear in the Send Fax dialog box by using the Search Records dialog box. When you fill in one or more field, the phonebook list will find and display all of the records in the current phonebook that match that field or those fields.

FIGURE 11.8

The Send Fax dialog box with Groups active

Delrina WinFax PRO Send

To : Allen, Barbara   << Phonebook   Add to Phonebook...

Number : 555-6543   Add to List

Phonebooks
Ballads

Subject :

Recipient List

Name   **Company**
Allen, Barbara
Henry, John   Big Bend Tunnel
Hamilton, Mary
Jones, Casey   Illinois Central Ra

Schedule...
Remove
Make Group...

Modify... Search... Restore...

Display
◉ Individuals   ○ Groups

☐ **Cover Page** Quick Cover Page   Fill In... Cover...

Attachments: 0 attachments selected   Attach...

Options: Fine Res, Preview   Options...

Fax: 555-6543
Voice:
Misc:
Notes: Scarlet Town Enterprises

Send   Cancel

There are several ways to use this feature. You can type a partial name in the Company Name field to limit the list to companies whose name includes that string—for example, if you type **Unit**, you might see "Unitarian Church," "Unity Bookshop," and "United States Coast Guard."

If you took our advice in Chapter 10 and used a standard format in the Notes and Misc fields to identify different classes of phonebook records, you can use an item in one of those fields as your search criterion.

Recipient List   Sometimes you may want to send the same fax to a list of names that you had not previously defined as a phonebook group. You can create an instant group by adding names to the recipient list. You can also use the recipient list to schedule a delayed transmission of a fax to one or more recipients.

To add a name to the recipient list, either type the name and fax number in the To and Fax fields or highlight that name in the phonebook list and then click on the Add To List button. Or drag the name from the phonebook list to the Recipient list. You can add a group to the recipient list by highlighting the name of the group and either clicking on Add To List or dragging the group name to the recipient list.

Remove   If you add a name to the recipient list by mistake, you can remove it by highlighting the name and clicking on the Remove button, or by dragging it outside of the recipient list box.

Make Group   After you create a temporary group in the recipient list, you can add that group to the current phonebook by clicking on the Make Group button. When the New Group dialog box shown in Figure 11.9 appears, type the name you want to assign to this group.

FIGURE 11.9

The New Group dialog box

Filling in the WinFax 4.0 Expanded Phonebook Section

In WinFax Pro 4.0, the Phonebook section of the Send Fax dialog box looks a little different from the earlier version, as shown in Figure 11.10.

Phonebooks   The Phonebooks field contains the name of the current phonebook. Use the arrow at the right side of the field to drop down a selector list if you want to display names from a different phonebook.

If there is not enough room for all the names in the current phonebook, use the up or down arrows to display additional names. When you select a name from the current phonebook, that name and fax number will appear in the To and Number fields, and selected information from that phonebook record will appear at the bottom of the expanded dialog box.

FIGURE 11.10

The WinFax Pro 4.0 Send Fax dialog box with the Phonebook section visible

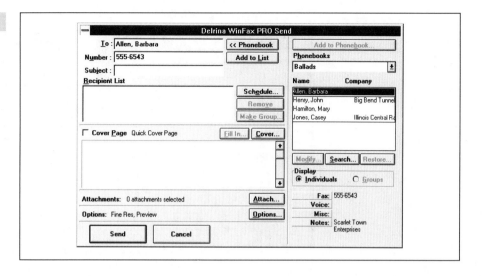

Add to Phonebook... Click on the Add to Phonebook... button to place a new record in the current phonebook.

Modify... Highlight a name and click on the Modify... button to change all or part of an existing phonebook record.

Search... Click on the Search... button to display the Search Records dialog box shown in Figure 11.11. Fill in one or more fields in the dialog box to search for phonebook records in the current phonebook with matching fields. For example, if you type a partial name in the Company Name field, the list will be limited to companies whose name includes that string—for example, if you type **Unit**, you might see "Unitarian Church," "Unity Bookshop," and "United States Coast Guard."

If you took our advice in Chapter 10 and used a standard format in the Notes and Misc fields to identify different classes of phonebook records, you can display a list of records that contain the same entry in one of those fields. Click on Restore... to see the entire phonebook again.

Display The default display shows individual phonebook records. Click on Groups to display a list of phonebook groups. The Send Fax dialog box will change to show a list of groups and the

FIGURE 11.11

The Search Records dialog box

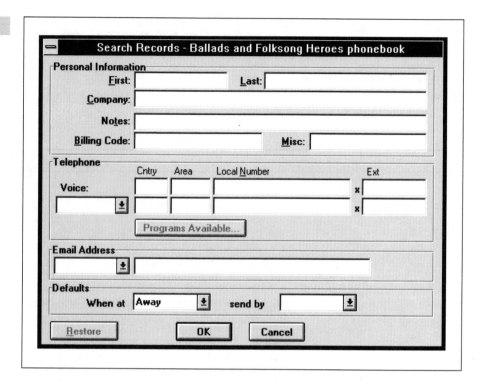

members of the current group, as shown in Figure 11.12. To send the same fax to all members of a group, highlight the group name. Click on Individuals to change the dialog box back to its original format.

Variable Information   In the lower right portion of the expanded Send Fax dialog box, there are four fields that can each display any field in the current phonebook record. To change a field, click on the field name and select a new item from the menu shown in Figure 11.13.

Scheduling Delayed Transmission

Normally, WinFax sends new faxes as soon as you enter the Send command, but there are several reasons why you might want to schedule a fax for delayed transmission. For example, you can save money by waiting until long-distance rates are cheapest. If you're sending something to another time zone, or if you know the recipient is away from the office, you can delay transmission until the time you expect the recipient to be there

FIGURE 11.12

The Send Fax dialog box with Groups active

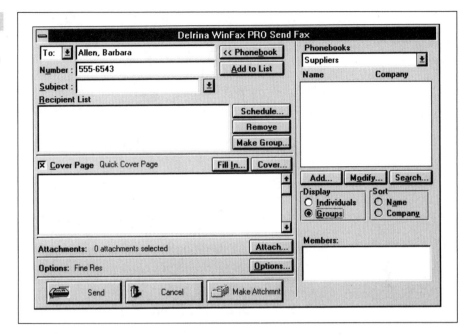

FIGURE 11.13

The Send Fax field name menu

to receive your fax. You can also use scheduled transmission to prepare an announcement in advance, but delay actually sending the faxes until after a specific date or time.

To schedule delayed transmission of a fax, highlight the name of the recipient or group in the recipient list and click on the Schedule button. When one of the Schedule dialog boxes shown in Figures 11.14 and 11.15 appears, click on the time or date box to change the setting. These fields won't let you type the time or date directly, so you must click on the up or down arrows to make the changes.

In WinFax Pro 4.0, you can display the calendar for the current month like in the one in Figure 11.16 by double-clicking the date field. To choose another date in the same month, click on the date. To choose a different month or year, use the arrows on the side of the year and month name fields.

You can schedule transmission to different names in the recipient list at different times by highlighting each name separately and clicking on the Schedule button. When you schedule a fax for delayed transmission in WinFax Pro 3.0, that fax appears in the WinFax event list (in the main WinFax screen) with an hourglass next to the listing. In WinFax Pro 4.0, there's a separate list of scheduled transmissions in the Outbox window.

FIGURE 11.14

The WinFax Pro 3.0 Schedule dialog box

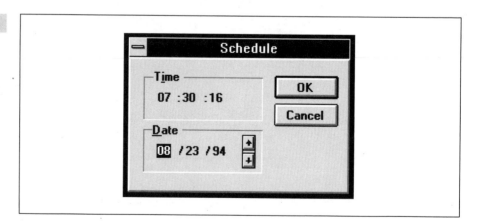

FIGURE 11.15

The WinFax Pro 4.0
Schedule/Modify
Events dialog box

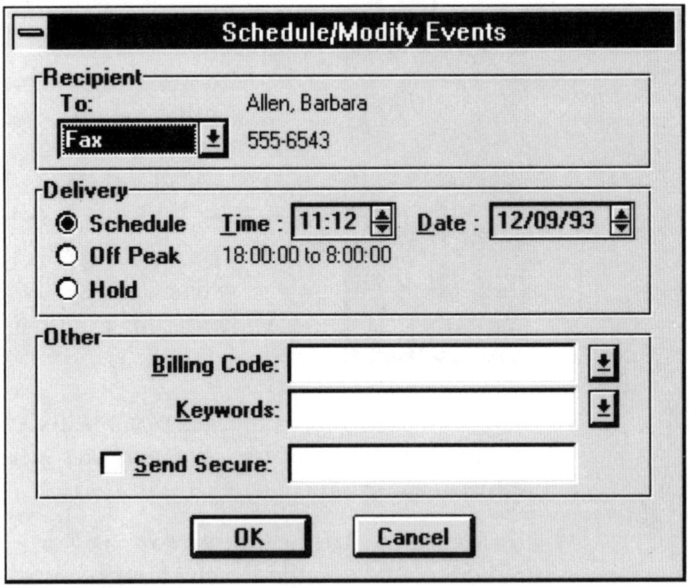

FIGURE 11.16

WinFax Pro 4.0
Calendar dialog box

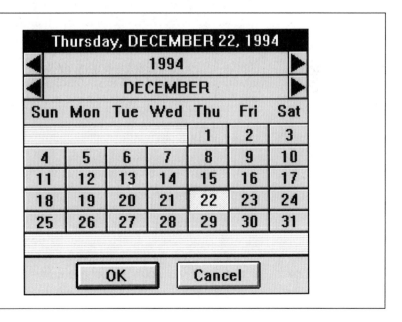

NOTE

It may seem obvious, but it's worth mentioning that the computer must be on and WinFax running at the scheduled transmission time. That's a good reason to move the WinFax icon to your Windows Startup group, so that the program runs automatically, every time you start Windows. It's okay to minimize the WinFax screen to an icon, but let it run in background all the time. If WinFax isn't running at a scheduled transmission time, it will automatically send the scheduled fax the next time you start the program.

If you change your mind after you schedule a fax transmission, you can modify the date or time or cancel the scheduled transmission from the WinFax screen.

To change the date or time in WinFax Pro 3.0, highlight the event in the list, and either click on the Resched button or select Reschedule from the Send menu. To cancel a scheduled transmission, highlight the event and either click on the Delete button, or select Remove Event from the Fax menu.

In WinFax Pro 4.0, use the Outbox command in the Fax menu to open the Scheduled Events screen, and use the Reschedule... command in the Outbox menu to display the Schedule dialog box.

Instead of scheduling transmission for a specific time and date, it's also possible in Version 4.0 to instruct WinFax to schedule transmission during "off-peak" hours, or to hold the fax for unspecified future disposal.

In WinFax Pro 4.0, you can also make several other changes to a scheduled fax transmission:

- If the person to whom this fax is directed is using a fax modem that supports binary file transfer or the Microsoft At Work architecture, you might be able to send this fax more quickly by choosing a different transmission method. Use the drop-down menu in the Version 4.0 Recipient box to choose fax, binary file transfer, Microsoft At Work, or MCI Mail as your transmission method. Chapter 16 contains more information about using alternative file types.

- If you have assigned a billing code to a phonebook entry, that code appears in the Other section of the Schedule/Modify Events dialog box. If you want to assign a different billing code to this transmission, type the new number in the Billing Code field.

- The Keywords field allows you to add a description of this transmission to the record of this call in the Send Log. If you want to make sure delivery of this fax or file is restricted to your intended recipient, make Send Secure active and type in the CSID code programmed into the receiving fax machine or modem. WinFax will not deliver this fax if it connects to a fax machine or modem whose CSID does not match the one specified.

Adding a Cover Page to a Fax

Cover pages usually contain the names and fax numbers of both the sender and intended recipient of a fax message, along with the total number of pages in the fax and the sender's voice telephone number. If you're faxing a brief message, you can frequently include it on the cover sheet and send it as a one-page transmission.

WinFax Pro includes a library of about a hundred special cover sheets, which you can send as-is or modify to meet your own special needs. You can also create your own custom cover sheets and add them to the library.

Use the options in the Cover Page section of the Send Fax dialog box to add a cover sheet to your fax:

Send (WinFax Pro 3.0 only) or **Cover Page** (WinFax Pro 4.0 only)   Place a checkmark next to Send or Cover Page to add a cover sheet to this transmission as the first page of your fax.

Select... (WinFax Pro 3.0 only) or **Cover** (WinFax Pro 4.0 only)   Click on the Select or Cover button to choose a different cover sheet from the Cover Page Library dialog box. We'll explain how to work with the Cover Page Library later in this chapter.

Fill In...   Many cover sheets have fields that can contain either standard text or specific text for this fax. When you click on the Fill In... button, the Cover Page Filler dialog box shown in Figure 11.17 appears, with the current cover page on display.

FIGURE 11.17

Cover Page Filler
dialog box

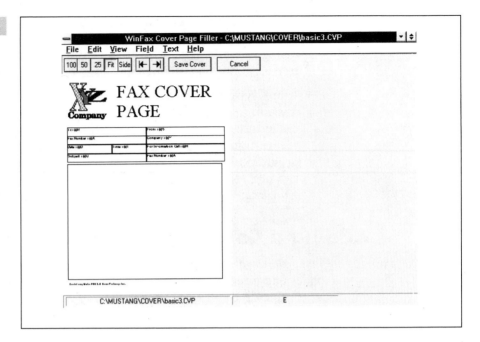

You can use the tab keys, the Field menu, or the tabbing buttons (the two buttons with arrows on them) in the button bar to move between text fields. When the cursor is in the text field where you want to place the text, type the text.

Text fields are not the same as variable blocks, but you can include variables in a text field. Variables are specific items of information (such as the name of the sender or today's date) that WinFax inserts in cover pages in place of a variable command. Look in the section on cover pages later in this chapter for more information about variables.

Creating Attachments

You send a fax through WinFax by entering a print command from a Windows application. Simple enough. But how do you combine information from more than one application into a single fax message? You could probably use a file format conversion utility that recognizes files from different applications, or you could send the output of each application separately, but WinFax includes an easier method: instead of printing directly to WinFax, you can save pages as attachment files.

To create an attachment file, print a document to WinFax as if you were sending it as a fax. But when the Send Fax dialog box appears, select Save To File… in WinFax Pro 3.0 or click on the Make Attchment button in Win-Fax Pro 4.0 to display the Save To File dialog box shown in Figure 11.18 if you have Version 3.0 or Figure 11.19 if you have Version 4.0.

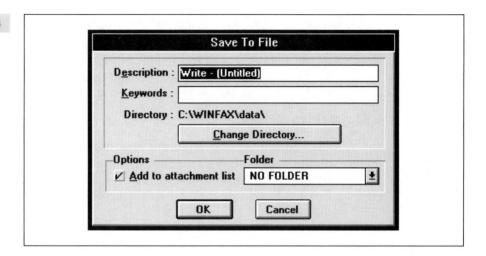

FIGURE 11.18

WinFax Pro 3.0 Save To File dialog box

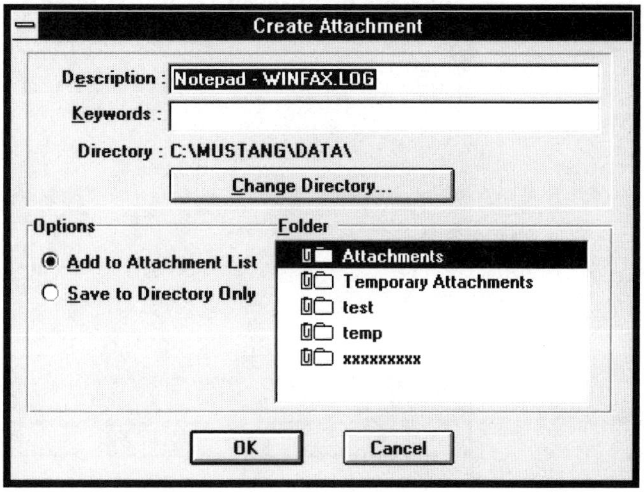

FIGURE 11.19

WinFax Pro 4.0 Create Attachment dialog box

When the Save To File dialog box appears, it will show the name of the original application in the Description field. You may want to edit this field to provide a more specific description of the document you're saving. If you expect to search for this file later, add search topics in the Keywords field. The default directory for attachment files is the DATA subdirectory under the WINFAX directory. To store the file in another location, use the Change Directory... button.

If the Add To Attachment List option is active, WinFax will place this document in the attachment list right away; if it is not active, you will have to add the document to the list before you use it. To add this attachment file to a folder, highlight the name of the folder in the list of folders.

Adding Attachments to a Fax

To add an existing attachment to a fax, select Add... in the Attachments box of the WinFax Pro 3.0 Send Fax dialog box to display the Add Attachments dialog box in Figure 11.20, or Attach... from the WinFax Pro 4.0 Send Fax dialog box to display the Select Attachments dialog

FIGURE 11.20

WinFax Pro 3.0 Add Attachments dialog box

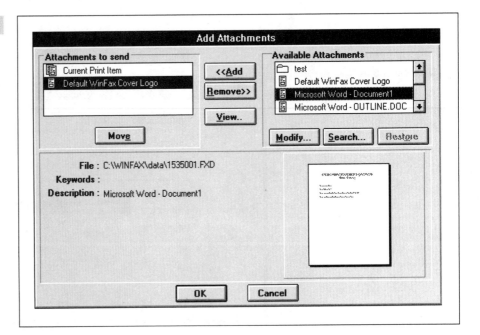

box in Figure 11.21. Select documents from the Available Attachments list in the Version 3.0 Add Attachments dialog box or from the Attachment Library list in the Version 4.0 Select Attachments dialog box by either highlighting the description and clicking on <<Add or Add To Send List, or by dragging the description to the Attachments To Send list. The bottom part of the dialog box contains more information about the currently highlighted attachment and a small picture of the document. To see the contents of a folder, double-click on the folder's description.

FIGURE 11.21

WinFax Pro 4.0 Select
Attachments dialog box

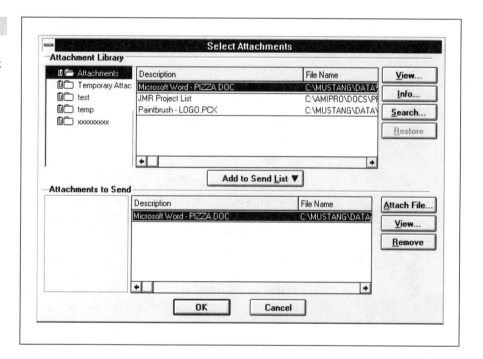

The Add Attachments and Select Attachments dialog boxes include these additional options:

Add (WinFax Pro 3.0 only)   Click on this option button to include the currently highlighted attachment or folder to the Attachments To Send list.

Attach File... (WinFax Pro 4.0 only)   Click on Attach File... to choose a file that is not already part of the Attachment Library list.

Remove   Click on this option button to delete a previously selected attachment from the Attachments To Send list.

View...   Use this option to start the Viewer program with the currently highlighted attachment displayed. The next chapter contains more information about the Viewer program. In WinFax Pro 4.0, the View... command also starts other programs that can display the currently selected file. For example, if the current file is a text file with a .DOC file extension, the View... command will open Word for Windows with the selected file.

Move (WinFax Pro 3.0 only)   Use Move to change the order in which WinFax will send attachments and folders. To move an item in the Attachments To Send list, highlight the item, click on the Move button and use the up and down keys to select another location in the list. When you're satisfied with the new location, click on the End Move button.

It's also possible, and probably easier, to change the order of the Attachments To Send list by dragging items up or down with your mouse.

Modify (WinFax Pro 3.0 only)   Select the Modify button to change the organization of your fax attachments and folders. Chapter 13 contains more detailed information about managing attachments and about how to use the Fax Attachments dialog box.

Info... (WinFax Pro 4.0 only)   Highlight an attachment description and click on the Info... button to see more information about that attachment, including the type of file, number of pages, file name, and a thumbnail view of the first page.

Search   The Search option in the Add Attachments dialog box works like the phonebook search function. When the Search Attachments dialog box appears, type either the description or one or more keywords. If Case Sensitive Search is active, the search will be limited to the exact upper- and lowercase form that you type. If Match Anywhere is active, you can search for matching character strings anywhere in the description or keywords. For example, if you type **unit** in the Description field, with Case Sensitive Search not active, WinFax might find both "United Popcorn memo" and "Shady Grove order—100 units."

It's a whale
of a job.
Here's our
estimate.

Restore   The Restore button is normally not available, but after you run a search, you can use it to restore the complete list of attachments.

Additional Send Fax Options

The Send Fax dialog box also includes a handful of options that you probably won't want to change very often. To display these items, click on the Options>> or Options... button. In WinFax Pro 3.0, the Options section of the Send Fax dialog box appears, as shown in Figure 11.22. Click on Options>>

FIGURE 11.22

The WinFax Pro 3.0 Send Fax dialog box with options visible

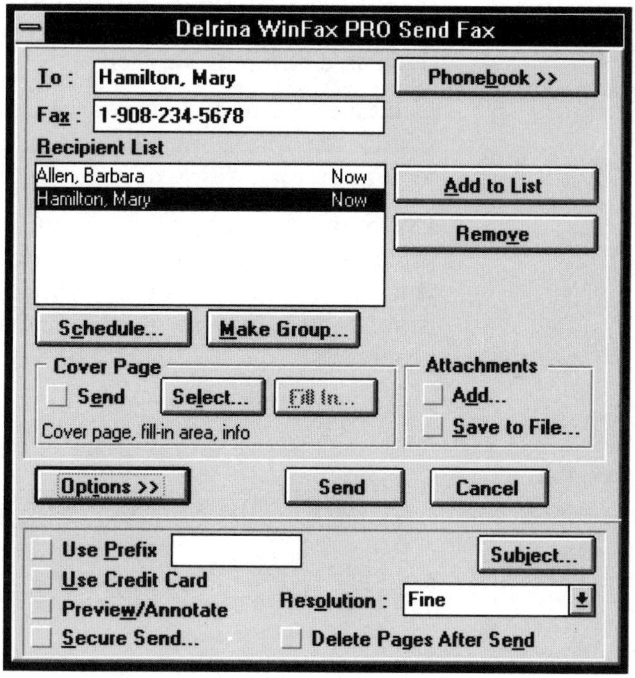

again to hide the options section of the dialog box. In WinFax Pro 4.0, the separate Send Options dialog box shown in Figure 11.23 appears.

The expanded Send Fax dialog box and the separate Send Options dialog box include these options:

Use Prefix or Use Dial Prefix   Make this option active to add or change a prefix at the beginning of the fax number that WinFax dials. If your telephone system requires you to dial 9 for an outside line, this is the place to add it. If you put a comma after the number, WinFax will pause after the 9 to wait for a second dial tone. To extend the duration of the pause, add additional commas.

The default prefix is the one you specified in the Dial Setup dialog box during Program Setup. In most cases, the only time you will need to change this prefix is if you're using a portable computer in a different location.

Use Credit Card   If you specified a credit card number in the Dial Setup dialog box, you can instruct WinFax to charge this call to your credit card by making this option active.

Preview/Annotate   Make this option active to start the Viewer program, which lets you see the document on your screen before you send it, and if you wish, add annotation to the document. The next chapter contains more information about using the Viewer program.

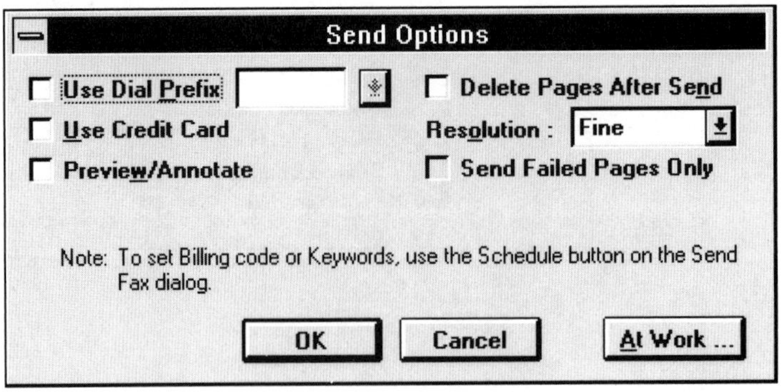

FIGURE 11.23

The WinFax Pro 4.0 Send Options dialog box.

Send Options dialog box showing: Use Dial Prefix, Use Credit Card, Preview/Annotate, Delete Pages After Send, Resolution: Fine, Send Failed Pages Only. Note: To set Billing code or Keywords, use the Schedule button on the Send Fax dialog. Buttons: OK, Cancel, At Work...

Secure Send... (WinFax Pro 3.0 only)   When this option is active, WinFax will not send this fax unless the recipient's fax machine or fax modem returns the station identifier you type in the Secure Send dialog box in Figure 11.24.

FIGURE 11.24

The Secure Send
dialog box

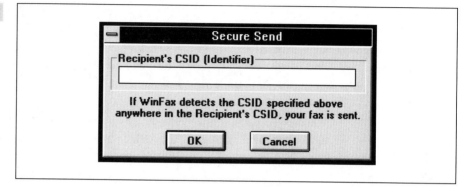

To use this feature, you must know the recipient's station identifier. The easiest way to find this information is to send the same recipient a fax that does not contain any confidential information, and look at the Identifier field in the Status dialog box (if Display Call Progress is active in the Program Setup dialog box), or the WinFax Send Log.

Subject   Select the Subject... button to assign a subject description, keywords, or a billing code to the record of this transmission in the Send Fax log. In the WinFax Pro 4.0 Send Fax dialog box, the Subject field is directly under the Number field.

Resolution   WinFax uses the default resolution unless you choose a different resolution in this field. You can change the default from the WinFax Setup dialog box. In most cases, you should use Fine resolution for the best possible image quality. But if you want to hold down your long-distance telephone charges, you can use Standard to reduce the amount of time needed to transmit each page.

Delete Pages After Send   When this option is active, WinFax will not save this document as page files after it successfully sends them. You can still re-send the document from the original Windows application, but if you included any attachments with this fax, you will have to re-assemble them from the attachment files. If you delete pages after sending them, the Send log will list the transmission, but it won't be possible to view it or re-send it.

Send Failed Pages Only (WinFax Pro 4.0 only)   Use this option after you try to send a fax but lose the connection after the first few pages. WinFax will re-send the current fax, starting with the first page that failed.

At Work... (WinFax Pro 4.0 only)   If you're sending this fax to a recipient whose system also supports the Microsoft At Work architecture, click the At Work... button to set your system to match the recipient's configuration.

Broadcast (WinFax Pro 4.0 only)   If you use Delrina's Fax Broadcast service, you can send the same fax to many recipients with a single telephone call. To broadcast this fax, select the Send Using Service option.

Keeping Track of Call Status

If Display Call Progress is active in the Program Setup dialog box, the status dialog box in Figure 11.25 appears on your screen whenever Win-Fax is sending a fax message, even if the WinFax program is minimized.

The Operation field in the Status dialog box shows the specific type of data that your computer is currently exchanging with the distant fax machine or modem. Unless you're having trouble making a connection, this is more information than you really need (which is why you don't have a similar status display on most fax machines). The Current Page field also

FIGURE 11.25

The Call Status
dialog box

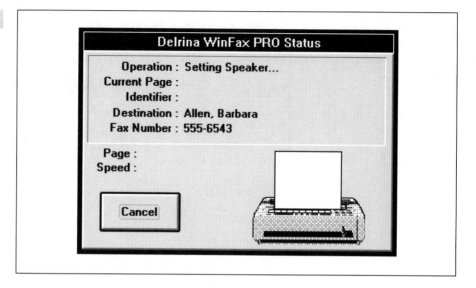

shows the percentage of each page that WinFax has transmitted, and the Page field shows the current page being transmitted and the total number of pages.

The Identifier field contains the station identifier (the CSID) that the distant fax machine or modem has returned to WinFax. The Destination and Fax Number come from the Send Fax dialog box.

It's not really necessary to display the Status dialog box when you send a fax, since the actual transmission takes place in background while you run other Windows applications. If Display Call Progress is not active in the Program Setup dialog box, you can check on the current status of a fax by looking at the WinFax program screen. The same status information that appears in the Status dialog box is visible in the message area of the WinFax screen, as shown in Figure 11.26.

FIGURE 11.26

WinFax Screen with an active status message

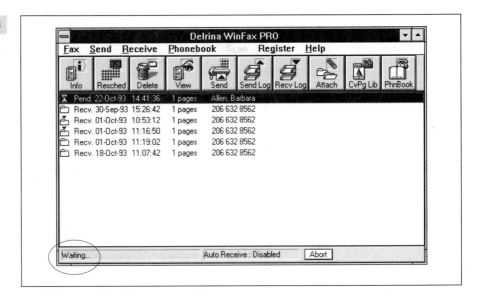

Working with Cover Sheets

Earlier in this chapter, you learned that you can add a cover sheet to a fax transmission from the Send Fax dialog box. In this section, we'll explain how to modify designs in the library of cover sheets and how to create a new cover sheet design.

The Cover Page Designer Program

The WinFax Pro package includes a library of about a hundred cover pages, which you can use in their original form, or modify to meet your own specific needs. To add one of these cover sheets to a fax in WinFax Pro 3.0, click on the Select... button in the Cover Page box of the Send Fax dialog box, and make Send active. In the WinFax Pro 4.0 Send dialog box, make Cover Page active and click on the Cover... button to display the Select Cover Page dialog box.

To change one of the standard cover pages in WinFax Pro 3.0, or to create a new one, start the Cover Page Designer program, from either the Windows Program Manager or by selecting the Design... button in the Cover Page Library dialog box.

In WinFax Pro 4.0, you can start the Cover Page Designer from the Windows Program Manager by clicking on the Cover Page Designer icon button. If WinFax is already running, select the Cover Pages command from the Fax menu or click the on the Cvr Pages Button and double-click on New Cover Page at the top of the list of cover pages, or use the New command in the Cover Pages menu. When the New Cover Page dialog box appears, select the Design... button.

When the Cover Page Designer screen in Figure 11.27 appears, you can select an existing cover sheet file by choosing File ➤ Open..., or create a new one by choosing File ➤ New... in the same menu.

FIGURE 11.27

The Cover Page
Designer screen

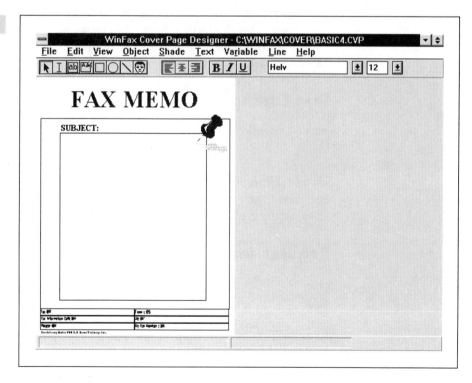

When you create a new cover sheet, you have the opportunity to specify the size of the page. While all the standard cover sheets are letter size (11 inches long), there really isn't any reason not to save transmission time and paper at the receiving end (especially if the recipient has a conventional fax machine with a roll of paper) by using a Half Page cover sheet, or even a three- or four-inch Custom sheet.

The Cover Page Designer program treats each block of text, space for fillable text, and graphic element as an *object*. You can use menu commands and commands in the button bar (described below) to add, remove, and change objects to a cover sheet.

Select

Before you make any changes to an object, you must select it. To select an object, click on the Select Tool icon, move your mouse to the object, and click on it. To select multiple objects, hold down the Shift key while you click on each object.

Text Editing

To add, change, or remove text in an object, select the Text Editing tool and move the cursor to the location where you want the text to appear, or to existing text you want to change or remove.

Regular Text

Regular Text is a permanent part of a cover page. When you define text as Regular Text, it will be part of the standard version of this cover sheet.

To create Regular text, select the Regular Text tool, move your mouse to one corner of the location where you want to place the object, hold down the left mouse button, and drag the pointer to the opposite corner. After you create an object, you can use the Select tool to change its size or position on the page.

Fillable Text

Fill Text is text that you enter for a specific fax cover page. When you define a Fillable Text object, you will be able to type a message within this object by using the Fill In... button in the Send Fax dialog box. To create a Fillable Text object, select the Fillable Text tool, move your mouse to one corner of the location where you want to place the object, hold down the left mouse button, and drag the pointer to the opposite corner. After you create an object, you can use the Select tool to change its size or position on the page.

Box, Oval, Line

The Box, Oval, and Line tools are all drawing tools. To add a box, oval, or line, select the appropriate button and use your mouse to draw the figure. To draw a perfect square, select Snap To Grid from the View menu and use the Box tool. For a circle, select Snap To Grid and use the Oval tool.

Graphic

The WinFax graphic tools are pretty primitive compared to more sophisticated drawing and painting programs like CorelDRAW. And sometimes you may want to use clip art or a scanned-in logo as part of your cover sheet. To import a graphic image into a cover sheet, select the Graphic tool and use your mouse to size and locate an object. Then either double-click on the new object or click on it once and select Object ➤ Graphic. When the Graphic Attributes dialog box shown in Figure 11.28 appears, either type the name of the graphic image file in the Filename field or click on the Select button to display a Select dialog box. In the Scaling box, choose None to maintain the original size and proportions, Fit In Window to vary the height and width of the image, or Aspect Fit to maintain the original proportions but vary in size with the size of the object.

FIGURE 11.28

The Graphic Attributes dialog box

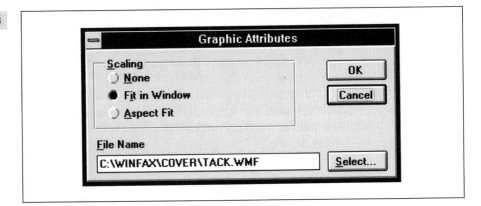

Text Controls

The remaining buttons in the button bar control the text alignment, type style, and font in cover page objects.

Menu Commands

Most of the commands in the Cover Page Designer menu are similar to the ones you probably recognize from other Windows applications and utilities. The best way to learn how to use them is to experiment. However, a few commands have very specific uses in the Cover Page Designer, so we'll describe them in detail here.

Tabbing Order Tabbing Order is the sequence in which the cursor moves from one fillable text field to the next when you press the Tab key. The default tabbing order is the order in which the fillable text objects were created. The name of the current field is visible in the bottom right corner of the Cover Page Designer screen.

To change tabbing order, select Tabbing Order... in the Object menu. When the Tabbing Order dialog box appears, it lists all fillable text objects in order, one object to a line. To change the current tabbing order, highlight the name of the object you want to move, and drag it up or down to the new position.

Variables Variables are items of information that WinFax changes in each cover page. For example, when you include the Recipient Name variable in a cover sheet, it appears in the text as @R. When WinFax sends this cover sheet, it takes information from either the To field in the Send Fax dialog box or the first and last name fields in a phonebook record and prints the recipient's name in place of the @R.

You can use variables two ways. Either create a new object for each variable, or include them in the text you type into a text object.

To insert a new variable into a cover page, place the cursor where you want the variable and choose the variable item from the Variable menu. If the cursor is not currently inside an object, the program will create one automatically.

If you need to include an @ symbol in your text, you must type the symbol twice (it will only appear once). For example, to include an electronic mail address on a cover page, you must type the address as **name@@well.com**.

After you modify an existing cover sheet or create a new one, you can add it to the library as a file with a .CVP extension. As you do in most other Windows applications, you can save the modified version of a cover sheet file under the old name by selecting File ➤ Save.... To save the file under a new file name, or in a different location on your disk (or on a different disk), use the Save As... option in the File menu.

In this chapter, we've explained how to send faxes from Windows applications by "printing" to WinFax, and how to send faxes directly from WinFax. Regardless of the source, you must use the Send Fax dialog box to specify the recipient, add a cover sheet and attachments, and control the way WinFax will send this fax.

In the next two chapters, we'll tell you how to display and print copies of fax messages that you have sent or received, and how to use attachment files, event records, and archive files.

CHAPTER

12

Viewing and Printing Faxes

WHEN WinFax sends or receives a fax, it stores a copy of each page as a graphic file. Using the Fax Viewer program, you can display these files on your computer's screen, make paper copies on your printer, or re-send them as faxes. This chapter contains detailed instructions for using the Fax Viewer program.

Starting the Fax Viewer Program

The Fax Viewer is a separate program from the main WinFax program, but WinFax uses the Viewer to display existing fax files. You can open the Viewer by selecting the Fax Viewer icon in the WinFax Pro group in the Windows Program Manager.

If you're already running WinFax, there are several ways to start the Viewer program. From the main WinFax screen, you can select View from the Fax menu or click on the View button in the button bar. In WinFax Pro 4.0, the View button is at the extreme right end of the button bar, so you may have to maximize the WinFax screen before you can see it. There are also View buttons in many WinFax dialog boxes, including the Send and Receive logs and the Fax Attachments dialog box in Version 3.0. When you're in the main WinFax screen or a dialog box that contains a list of fax events, you can highlight an item on the list and click on the View button to start the viewer with the first page of the highlighted fax in the display.

There's one more way to start the WinFax Viewer. If the View Fax option in the Receive Setup dialog box is active, WinFax will start the Viewer program shown in Figure 12.1 automatically whenever it receives a new fax.

FIGURE 12.1

The Fax Viewer screen

Working with the Viewer Program

The WinFax Viewer program includes a standard Windows menu and a button bar with icon button commands that control the appearance of the current display.

Displaying an Image

If you started the Viewer from WinFax, or if the program started automatically because WinFax has just received a fax, the Viewer opens with a page in the display. When you start the Viewer from the Program Manager, the Viewer starts with no image in the display.

Once the Viewer program has started, you can display a new fax image by selecting File ➤ Open. When the file selector appears, either type the name of the field you want to display in the File Name field or choose the file from the list.

Moving around a Fax Message

If a fax message has more than one page, you can use the Next Page and Previous Page commands in the Page menu and the next and previous page buttons in the button bar to move backward and forward through the document. If the document has a large number of pages, you can use the Go To Page... command to move to any page.

Converting an Image to a Graphic Format

If you're using a graphics program or an application such as a word processor or desktop publisher that uses graphic images, you can export a WinFax image to a graphic format that the other program will recognize. Specifically, WinFax can export to Paintbrush (.PCX), TIFF, and Windows Bitmap (.BMP) formats.

To convert a fax image, select File ➤ Export. In the Export dialog box (shown in Figure 12.2), you can specify the format, file name, and path of the converted file. The file prefix must be no more than five characters, because the export utility saves each page as a separate file, with a three-digit number as the last three characters of the eight-character file name. If you only want to export part of a multi-page document, you can specify the particular pages you want to convert.

If you have added annotations to this fax, you can save the annotated version of the image by making the Merge Annotation option active. We'll talk more about annotation later in this chapter. If you expect to send the annotated version as a fax, save it in WinFax attachment format.

FIGURE 12.2

The Export dialog box

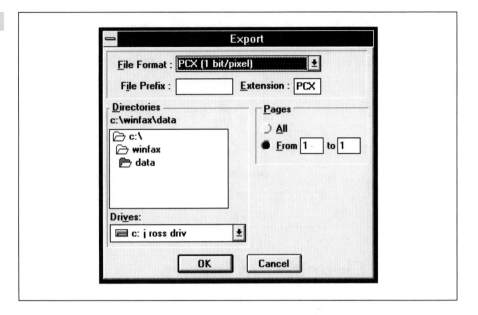

CAS fax modems use .PCX format as their default. You can either use the Export command to convert these .PCX files to WinFax format in order to add annotations, or use the OCR function to convert a fax to an editable text file.

There's another way to copy a WinFax graphic image to a Windows application—you can use the Windows Clipboard. Selecting Edit ➤ Select (or Select Area) allows you to select all or part of a page. The Copy command places the selected image on the Clipboard. When you move to another application, you can paste or copy from the Clipboard to import the WinFax image.

Printing a Fax Image

To print a fax with any changes you've made to the image in the Viewer, select File ➤ Print. You can change printers by either selecting File ➤ Printer Setup or by clicking on the Printer button in the Print Page(s) dialog box, shown in Figure 12.3.

FIGURE 12.3

The Print Page(s)
dialog box

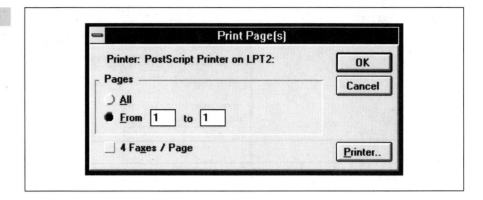

For miniature copies of a multiple-page fax, select the 4 Faxes/Page option.

Cleaning an Image

Noisy telephone lines sometimes add random black marks to a fax. You can use the Cleanup Fax feature to remove them from a WinFax image. These marks will be especially prominent when you view the image at a high magnification level.

WinFax Pro 3.0 Cleanup

The Cleanup Fax submenu in WinFax Pro 3.0 lets you choose to clean the entire page, clean just the portion of the page currently on your screen, or undo a cleanup before you save it.

Use the Save Cleaned Fax command to store an image after you have cleaned it. You can't undo a cleanup after you save it.

If the original fax is lighter than normal, the Cleanup function may remove parts of the original image. If this happens, use the Undo Cleanup command to restore the image to the condition in which you received it. To modify the extent of the cleanup process, add this line to the [Viewer] section of your WINFAX.INI file:

    FilterThreshold=5

A threshold level of 6 produces maximum cleanup. To reduce the amount of cleaning, reduce the Filter Threshold value.

WinFax Pro 4.0 Cleanup

The Cleanup Fax function is easier to use in WinFax Pro 4.0. When you select Edit ➤ Cleanup…, the Cleanup Fax dialog box shown in Figure 12.4 appears. You can specify the portion of the fax you want to clean and the amount of cleaning you want to do. A heavy degree produces the greatest amount of cleaning, and a light degree the least.

FIGURE 12.4

WinFax Pro 4.0 Cleanup Fax dialog box

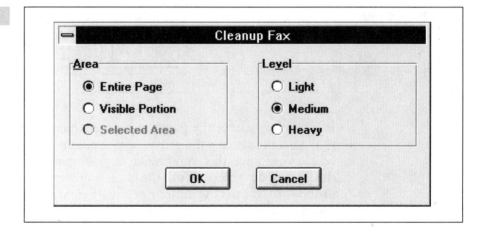

Altering the Image on Your Screen

The Viewer program displays new faxes oriented the way WinFax receives them—if the pages were sent sideways or upside down, that's the way you will see them on your screen. When you receive a paper fax upside down, it's no big deal to turn it around, but rotating a monitor is a real nuisance. Therefore, the Viewer program includes tools for rotating the image without the need to wrestle with your video display.

In some cases, it's more important to see the layout of an entire page than to read the text or see all the details of graphics. The Viewer program allows you to increase and reduce the size of the image and to superimpose a miniature view of the whole page over a close-up of a segment of the same page. You can also display all pages of a multiple-page fax at one time.

When you receive a fax of a photograph or other gray-scale image, the image on your screen may not have enough contrast. Selecting View ➤ Enhance View can improve the amount of detail in the display.

Rotating the Image

To change the orientation of an image, use the Rotate Current Page submenu in the Options menu, or select one of the rotate buttons in the button bar shown in Figure 12.5. In WinFax Pro 4.0, there's also a Rotate All Pages submenu.

FIGURE 12.5

Button bar rotate icons

If somebody places the pages in a fax machine upside-down when they send you a fax, that's the way you will receive them. Since WinFax prints faxes as graphics, this won't make any difference unless you want to read the fax on your screen or use the OCR feature to export the text to a word processor. To turn the entire document right side up with a single command, choose Options ➤ Rotate All Pages.

Figure 12.6 shows the Viewer screen with a sideways image. Figure 12.7 shows the same image after it has been rotated 90 degrees clockwise.

Changing the Size of the Image

The View menu contains five commands that alter the size of the image in the Viewer display. The same commands appear in the button bar. The 100%, 50%, and 25% magnification levels shown in Figures 12.8–12.10 are arbitrary numbers rather than exact proportions of the size of a printed page. However, they do give you an accurate description of the relative size of the three levels—an image with a 100% zoom factor is four times bigger than an image with a 25% zoom factor.

FIGURE 12.6

A sideways image, as
received by WinFax

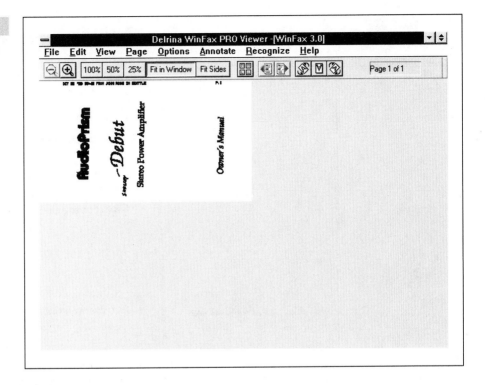

There are two other magnification settings that might be more convenient
to work with. Fit Sides, shown in Figure 12.11, makes the width of the
image equal the width of the Viewer window, and Fit In Window, shown
in Figure 12.12 displays the entire page.

If you're not sure which zoom factor you need, you can use the two magnifying glass icon buttons in Figure 12.13 to increase or reduce the size
of the image. When the Viewer is using the largest possible image, the plus
button doesn't work; when the image is already at its smallest, the Viewer
won't accept the minus button.

Besides the Fit In Window command, there are two other ways to display
a whole page in the Viewer. The All Pages command in the View menu and
the View All Pages button with four small pages on it display thumbnail
views of up to ten pages of a multiple-page fax at one time (as shown in
Figure 12.14).

FIGURE 12.7

The same image, after using the Rotate Current Page command

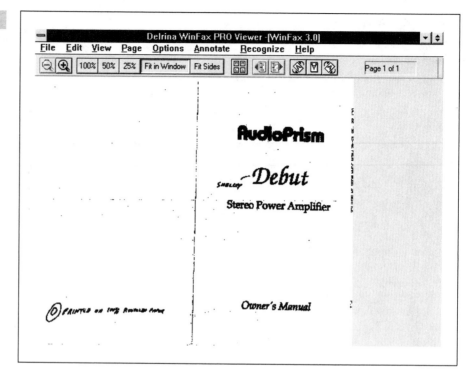

The Show Miniature command creates a window within the viewer screen that shows all of the current page. There's a box in the miniature view that shows the location of the larger image. If you drag your mouse inside that box, you will move it to a different segment of the page and drag the larger view along with it.

When the miniature view is active, the Show Miniature command in the View menu changes to Hide Miniature. To make the miniature view disappear, either select the Hide Miniature command or double-click on the button in the upper-left corner of the Miniature View window.

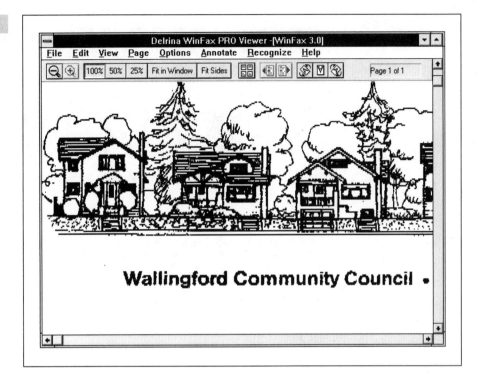

Enhancing the Image

The Enhance View command in the View menu increases the amount of contrast in the Viewer's display of a photograph or other gray scale image. This command changes the way an image appears on the screen, but it does not affect the quality of a printed copy of the image. If you're planning to export a received photo to another application, you will probably get a better result if you use a specialized tool, such as Aldus Photostyler or Adobe Photoshop, to improve the quality of the image.

FIGURE 12.9

The same image
at 50%

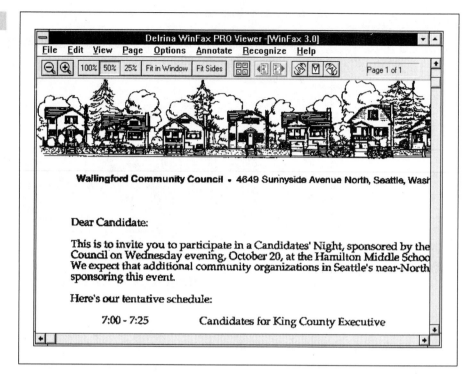

Reversing Black and White

In WinFax Pro 4.0, selecting View ➤ Invert (or Insert Display) displays an image like a photographic negative—white where the original was black, and black where it was white.

Other Things You Can Do from the Viewer

Besides displaying and printing faxes, the Viewer has two other major features: annotation and text recognition. When you receive a fax, it's sometimes useful to add your own comments to it before you save it, print it, or send it as a fax to somebody else. WinFax calls this *annotation*. When

FIGURE 12.10

The same image
at 25%

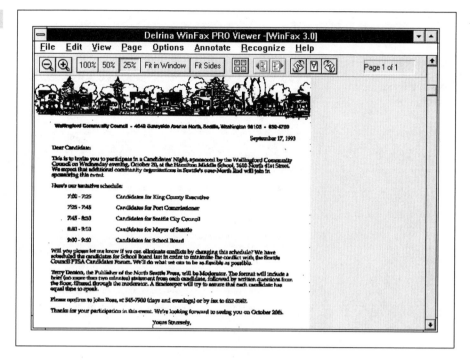

you receive information in a fax that you want to export to a word processor or other text-based application, the Viewer's text recognition function uses optical character recognition to convert received images to ASCII text. We'll explain both of these features in Chapter 15, "Advanced WinFax Pro Functions."

In WinFax Pro 4.0, you can discard any changes you make with the Viewer and return to the original image by selecting File ➤ Revert... before you save the altered image.

FIGURE 12.11

The same image set to Fit Sides

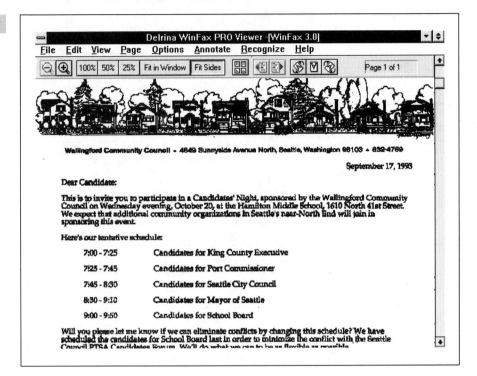

FIGURE 12.12

The same image at Fit
In Windows

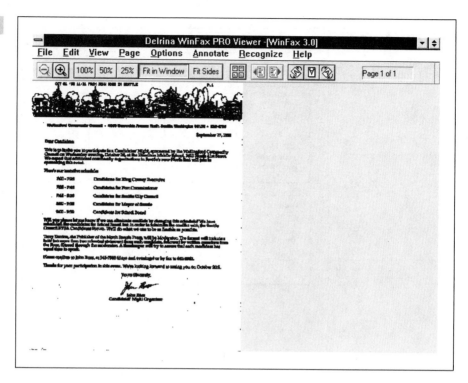

FIGURE 12.13

Magnifier button icons

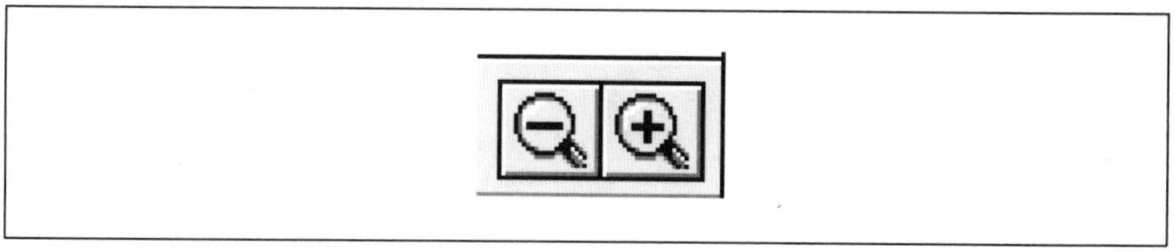

FIGURE 12.14

Pages shown in Multiple page view

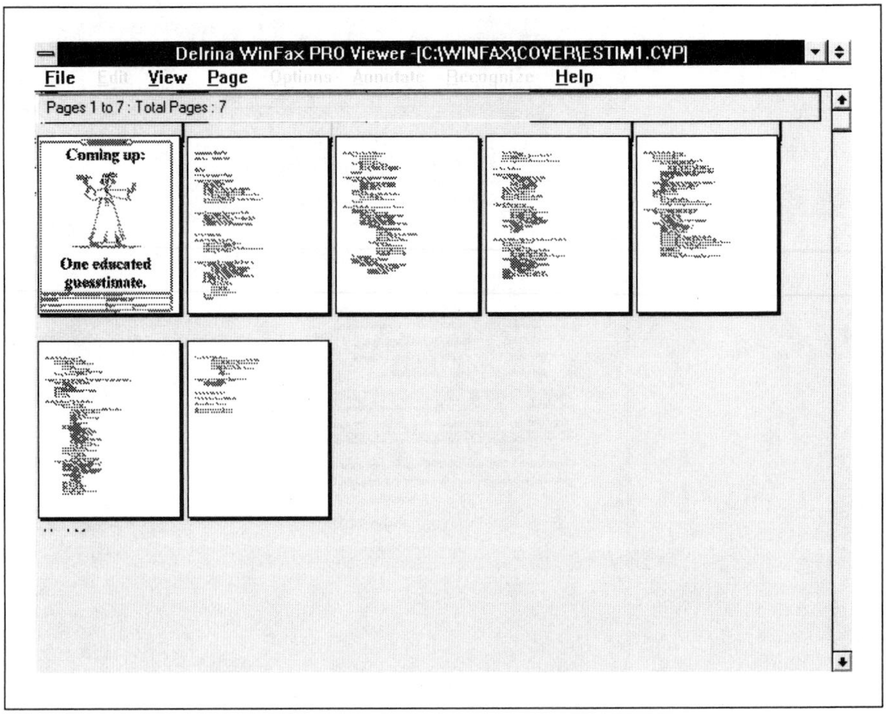

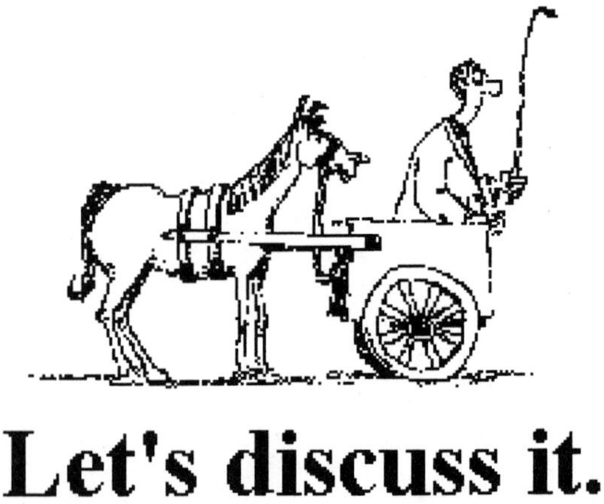

CHAPTER

13

WinFax Pro Fax Management

E VERY time WinFax Pro 3.0 sends or receives a fax, it treats each page as a separate fax image, which you can print, store on your computer's disk, change, re-send as a fax by itself, or attach to another document as a fax attachment. Each fax message that WinFax sends or receives and each unsuccessful attempt to send or receive a message is a "fax event." In WinFax Pro 4.0, you also have the option of storing multiple-page faxes.

In this chapter, we'll explain how WinFax works with fax images and fax events and how to use fax attachment files, event lists, send and receive logs, and archive files.

Using Fax Attachment Files

As you know, you can send a fax from almost any Windows application by using WinFax as your printer. But sometimes you may want to send pages from more than one application as part of the same fax transmission. That's where *fax attachments* come in. Instead of sending pages directly from an application to WinFax, you can store fax-ready pages as attachments, and add (or *attach*) them to other documents. We explained this process under the heading "Creating or Adding Attachments" in Chapter 11.

There are other ways to create attachment files. You can use the Scan And File... command in the Scan menu to save an image from a scanner, and you can use the Export command in the Viewer to convert pages from faxes that WinFax has sent or received to attachment file format (with .FXS file extensions).

It's also possible to use fax image files stored as .FXD files and .FXR files as attachments without converting them. WinFax creates an .FXD file for each page when it sends a fax, and creates an .FXR file for each page when it receives a fax.

Selecting Attachment Files

WinFax organizes attachments in a database. You can search for attachments and display thumbnail views by selecting the Attach button or the Attachments... command in the Fax menu to display the Fax Attachments dialog box shown in Figure 13.1.

Working with Attachments in WinFax Pro 3.0

The Attachments list in the upper-left corner of the dialog box contains a description of all the attachment files currently in the attachment list.

The Fax Attachments dialog box

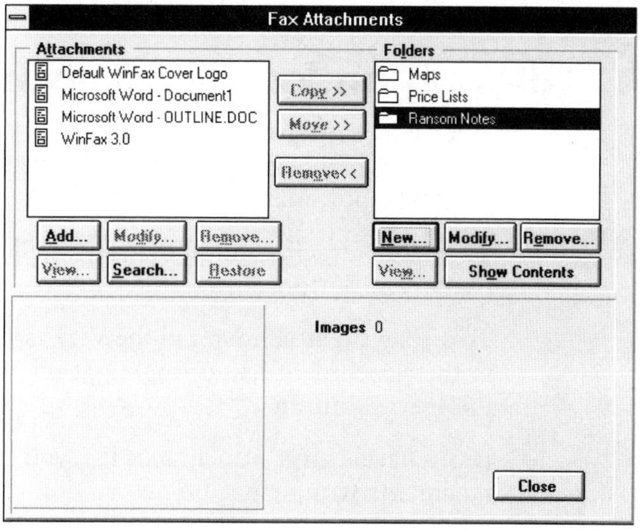

To display more information and a thumbnail view of an attachment in the lower half of the dialog box, select a description from the list.

Use the Add... button to add a new attachment file to the list. You can use the Add Attachment dialog box shown in Figure 13.2 to assign a description and keywords to fax images from faxes that WinFax has sent or received.

FIGURE 13.2

The Add Attachment
dialog box

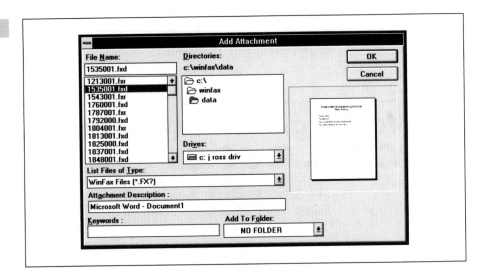

Altering an Existing Attachment File

To change the name, description, or keywords assigned to an existing attachment, highlight the description in the attachment list and select the Modify... button.

To take an attachment off the list, highlight the description and select the Remove... button.

To display an attachment in the Viewer program, double-click on a listing in the attachment list, or highlight the description of that file and select the View... button.

If you have a large attachment list, you can search through the list for attachments with a specific description or keyword (or both) by selecting the Search... button. Use the Restore button to return to the top of the attachment list.

Working with Attachment Folders

It can be useful to organize attachment files into groups, which WinFax calls *folders*. For example, you might want to store all your price sheets in one folder. When you want to attach one to a fax, you can search the Price sheets folder rather than the larger list of attachments, which might also include things like maps, product data sheets, and ransom notes.

The folders list in the upper-right corner of the Fax Attachments dialog box shows all existing folders. Click on the Show Contents button to list descriptions of all the attachment files in each folder under the name of the folder. Click on Hide Contents to return to a list of folders alone. To display the contents of a single folder, double-click on the name of that folder.

To add an attachment to an existing folder, highlight the attachment description in the Attachments list and use either the Copy>> button or the Move>> button. The Copy>> button adds an attachment to a folder without removing it from the Attachments list. The Move>> button places an attachment in a file and removes it from the Attachments list. If you prefer, you can use your mouse to drag an attachment description from the Attachments list to a folder name in the Folders list.

To remove an attachment from a folder, use the Show Contents button to display a list of attachments in each folder, highlight the attachment you want to remove, and click on the Remove<< button. If it's not already there, the attachment you remove will appear in the Attachments list.

To create a new folder, select the New... button under the folders list. Use the Modify... button to change the name of the currently highlighted folder. To delete a folder, select the Remove... button.

The View... button under the list of folders starts the Viewer with the first page of the first attachment in the folder on display. Use the page forward and page back icon buttons to move through other pages of the attachments in the folder.

Working with Attachments in WinFax Pro 4.0

The Attachments window contains a description of all the attachment files in the current attachment folder. To display more information about an attachment in the lower half of the window, select a description from the list and select Window ➤ Display Information, or move the cursor to the lower portion of the window and press the right mouse button. To display a small view of the attachment, select Window ➤ Display Thumbnails. Select Display View for a larger image.

Use the New button, select the New command from the Attachments menu, or double-click on New Attachment dialog box shown in Figure 13.3 to add a new attachment file to the list. You can use the New Attachment dialog box to assign a description and keywords to fax images from faxes that WinFax has sent or received.

To change the name, description, or keywords assigned to an existing attachment, highlight the description in the attachment list and select either the Modify button or the Modify command in the Attachments menu.

To take an attachment off the list, highlight the description and select the Remove button or the Remove command in the Attachments menu.

To display an attachment in the Viewer program, double-click on a listing in the attachment list, highlight the description of that file and select the View button or the View... command in the Fax menu.

FIGURE 13.3

The WinFax Pro 4.0 New Attachment dialog box

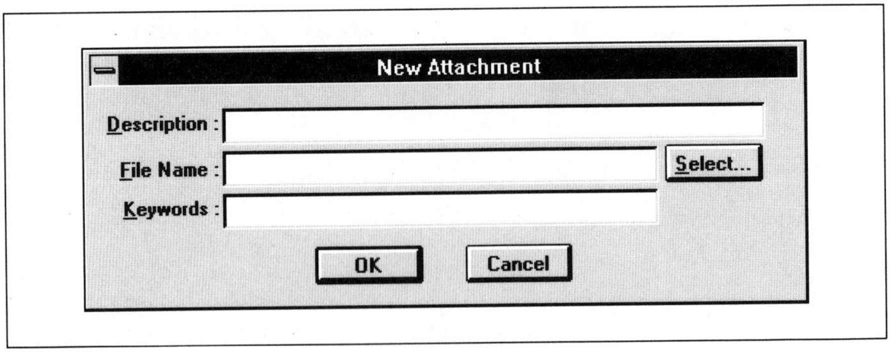

If you have a large attachment list, you can limit the list of attachments to those with a specific description or keyword (or both) by selecting the Search… command in the Attachments menu. Select Attachments ➤ Restore… or the Restore button in the Search Attachments dialog box to return to the top of the attachment list.

It can be useful to organize attachment files into groups, which WinFax calls *folders*. For example, you might want to store all your price sheets in one folder. When you want to attach one to a fax, you can search the "Price sheets" folder rather than the larger list of attachments that might also include other things.

The list in the left side of the Attachments window shows all existing folders. Click on the name of a folder to display descriptions of all the attachment files in that folder.

To move an attachment from the current folder to another existing folder, highlight the description in the Attachments list and use your mouse to drag an attachment description from the list to a folder name in the Folders list.

To remove an attachment from a folder, highlight the attachment you want to remove, and click on the Remove button or the Remove command in the Attachments menu.

To create a new folder, highlight New Folder in the list of folders, and select the New… button. Use the Modify button or the Modify command in the Attachments menu to change the name of the currently highlighted folder. To delete a folder, select the Remove button, or the Remove… command in the Attachments menu.

The View button and the View… command in the Fax menu both start the Viewer with the highlighted attachment on display.

Working with Fax Event Records in WinFax Pro 3.0

WinFax stores a copy of every fax it sends and receives as a fax image file. The event list in the main WinFax screen includes all outgoing faxes that have not yet been transmitted and all incoming faxes (including unsuccessful attempts) until you remove them from the list. The Send Log and Receive Log contain information about older fax events.

The WinFax Event List

When you select a fax event from the event list, you can use button bar commands to display the fax on your screen, print copies, and display a dialog box with more information. Figure 13.4 shows a WinFax event list.

Each item in the event list has an icon next to it that identifies its type: an hourglass for outgoing faxes that have not yet been transmitted or a file folder for received faxes.

To display more information about a fax event, highlight the listing and click on the Info button. A Pending Event Information or Receive Event Information dialog box appears as shown in Figures 13.5 and 13.6, with more details about the event than appear in the list itself. To start the Fax Viewer program with the first page of this fax in the display, select the View... button. To see thumbnail pictures of several pages at one time, select the Pages... button. To create or change the subject description, keywords, or billing code for this fax, select the Subject... button.

FIGURE 13.4

The Main WinFax Screen with several fax events

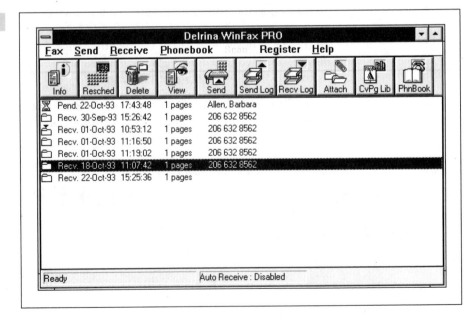

FIGURE 13.5

The Pending Event
Information dialog box

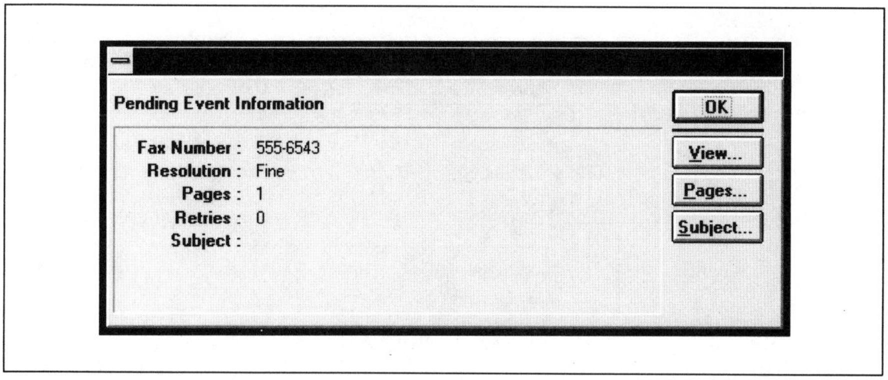

FIGURE 13.6

The Receive Event
Information dialog box

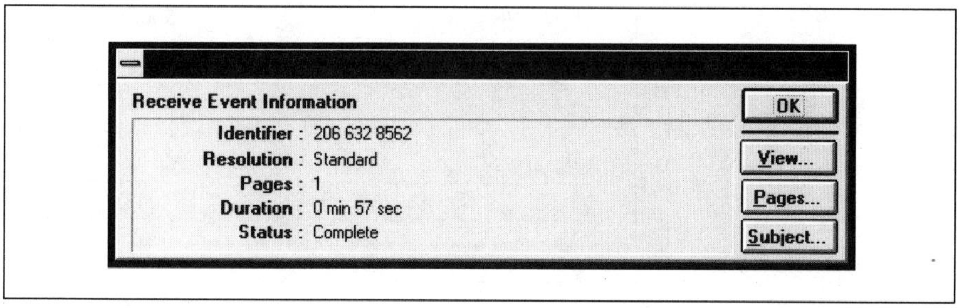

To remove an event from the event list, highlight the event listing and click
on the Delete command button. When you delete a pending fax event,
you cancel the scheduled transmission, but the event still appears in the
Send Log.

The Send and Receive Logs

The event list shows scheduled fax transmissions that haven't been transmitted yet and recently received faxes that you haven't removed from the
list. But what about older faxes? You can use the Send Log and Receive
Log to retrieve copies of old faxes, display information about each fax
event, and send another copy of the same fax.

To open a log, select the Send Log or Recv Log command buttons, or use
the Log command in the Send or Receive menus. As shown in Figures 13.7 and 13.8, both logs include lists of events and a block of expanded information about the currently highlighted event.

FIGURE 13.7

The Send Log

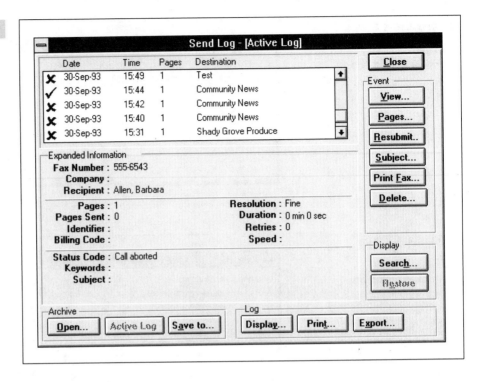

A check mark next to an event listing indicates that the event was successful—all pages of this fax were transmitted or received. An X indicates that the transmission was not complete. You can find the reason for the failure in the Status field.

Expanded Information

The Expanded Information block contains detailed information about one fax event. To display information about an event, select the listing of that event in the log.

Most of the Expanded Information fields are pretty clear. In the Send Log, the Pages field shows the total number of pages in the document, even if they didn't all get transmitted. The Pages Sent field shows the number of pages that actually made it to the recipient. These two numbers are equal when the transmission was successful.

FIGURE 13.8

The Receive Log

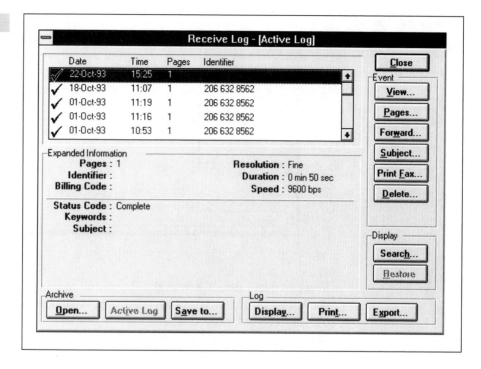

If WinFax receives a busy signal when it places a call, it will redial until it makes a connection. The Retries field in the Send log shows the number of times WinFax attempted to make the connection. You can configure this feature in the Dial Setup dialog box, which is buried under the Program Setup dialog box (in the Fax menu).

Event

Use the command buttons in the Event box to work with the currently highlighted fax event. The View button starts the Viewer program with the first page of the highlighted event in the display. If you're not already familiar with the Viewer, take another look at Chapter 12.

For thumbnail views of all the pages in the highlighted event (as shown in Figure 13.9), select the Pages button. If you have assigned a subject description to this event, it will appear under each page.

FIGURE 13.9

The Pages dialog box

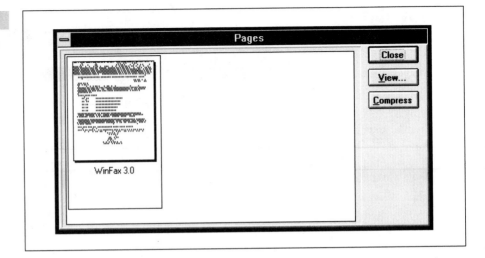

You can move directly from the Pages dialog box to the viewer by selecting the View... button. To compress the current image, click on the Compress button. A compressed image uses less space on your disk, but it takes longer to open. File Compression doesn't work if you turn off automatic compression in the WINFAX.INI file.

Use the Resubmit button in the Send Log to send another copy of a fax. When the Send Fax dialog box appears, the name and fax number of the original recipient is already filled in. To send the fax to a different recipient, type a new name and fax number.

Resubmit can be handy when you want to try to re-send a fax that failed the first time. If you select Send Failed Pages, WinFax will only send the pages that did not make it through before.

The Forward button in the Receive Log is similar to Resubmit in the Send Log—it's a method for sending a fax directly from WinFax. When you receive a fax that you want to send on to somebody else, Forward is the easiest way to do it. You can either send the whole fax, or just send selected pages. Figure 13.10 shows the Forward dialog box.

The Forward dialog box contains thumbnail views of all the pages in the fax event that was highlighted when you selected the dialog box. If you select the All Pages option, WinFax will forward the whole fax, including the original cover sheet (you can add your own cover sheet as a new first page from the Send Fax dialog box). To choose an individual page, make sure All Pages is not active and select the page you want to send. For more

FIGURE 13.10

The Forward dialog box

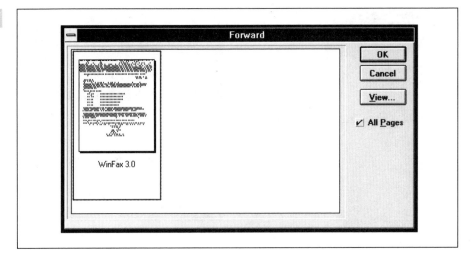

than one page, select the first page, then hold down the Shift key and click on the last page you want to send. For pages that were not grouped together in the original fax, hold down the Ctrl key when you select each page. To examine a page more closely, select that page and click on the View... button.

The Subject button opens the Subject dialog box for the highlighted fax event. Use this command to create or edit the subject description, keywords, or a billing code.

Select the Print Fax button to print a copy of the highlighted fax.

Some faxes are not important enough to save. If somebody faxes you a note that says "what are you doing for lunch today?" there's not a lot of value in storing it for future reference. Other faxes may be useful for a while, but not forever. Do you really need to save those weekly meeting agendas from last August? To delete items in the logs, select the Delete button. The Delete dialog box show in Figure 13.11 will appear.

The Delete dialog box in the Send Log offers a choice of all events in the log, or the currently highlighted events. If you want to delete several faxes at one time (but not all of them), hold down the Shift key when you click on each item that you want to delete. If you highlight an item by mistake, the same Shift-click routine deselects it. When Delete Pages (Keep Event) is active, the log will retain a record of any fax events you delete, but it

FIGURE 13.11

The Delete dialog box

will delete the actual images. When Delete Attachments is active, you will delete the attachment files (with .FXS file extensions), but not the send files (with .FXD file extensions).

If you select an item in the Send Log that was originally created in WinFax Lite or WinFax Pro 2.0, a Cover button appears in the Event box. Use the Cover button to display the View Cover Page dialog box.

If Recognize On Receive and Store Text In Receive Log were both active in the Receive Setup dialog box when WinFax received a fax, the Text button will appear in the Receive Log dialog box when you select that fax. Select the Text button to start the Viewer with the text in the bottom half of the screen.

Display

The buttons in the Display box let you instruct WinFax to limit the number of fax events in the event list to those that meet certain criteria. Choose the Search button to specify the criteria in the Search Log dialog box shown in Figure 13.12.

The log will display all events that match the information in the Search Log dialog box fields. If you type a date in the Events Before field, the log will only include the events that were sent or received before that date. If there's a date in the Events After field, the log will show events after that date. To search for events that occurred during a range of dates, put the starting date in Events After and the ending date in Events Before.

FIGURE 13.12

The Search Log
dialog box

Search Log

Events **B**efore :

Events **A**fter :

Identifier :

Subject :

Keywords :

Billing **C**ode :

Sta**t**us : | All

OK   Cancel

The Search Log is pretty forgiving about date formats. It will accept the current Windows date format (probably mm/dd/yy in North America), but it also recognizes 1994 instead of 94 as the year, and 2/5/94 instead of 02/05/94. If you're specifying the current year, you can leave out the year entirely, unless you're using the format yy/mm/dd. You can also spell out or abbreviate the month (for instance, *Oct 2* or *2 October*).

As an alternative, you can specify the number of days before today's date. If today is May 5, you could type **-3** in the Events Before field to display events up to May 2.

To limit the Send Log display to faxes for a single recipient, type that person's name (as it appeared in the To field of the Send Fax dialog box) in the Destination field.

If you have assigned a subject description, keywords, or a billing code to each fax, you can use these fields as search criteria. The Status field offers three choices—All, Completed, or Failed.

When WinFax displays the result of a search, the title bar of the Send Log or Receive Log shows the number of fax events that meet the search criteria and the total number of events in the log.

To return to a display of all events in the log, click on the Restore button.

Archive

Old fax event records fall into three broad categories: faxes that you expect to use again, faxes that you want to save because you might need them some day, and faxes that you don't care about. You can store the first type in the Send and Receive logs, and completely delete the junk faxes from your disk. The best way to handle the remaining events is to store them in a compressed archive, either on your hard drive or off-line on a floppy disk. By removing items from the Send and Receive logs and moving them to an archive, you can hold the size of the event lists in the logs to a manageable level.

To transfer a fax event from the Send or Receive logs to an archive, use the Save To... button in the Archive box. Use the Save To Archive dialog box shown in Figure 13.13 to specify the name of the archive file and the records that you want to archive.

The archive file is normally located in the DATA subdirectory under the directory that contains the WinFax program files. To place the archive file somewhere else on your hard drive or on a floppy disk, use the Select button to open up a file selector dialog box.

Use the options in the Records box to specify the fax records that you want to include in this archive. To remove events from the Send or Receive log when you put them in the archive, select the Delete Log Events From Active Log option.

FIGURE 13.13

The Save To Archive dialog box

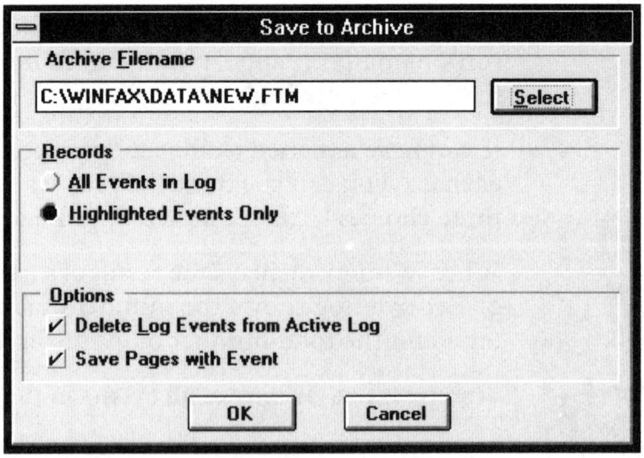

If you want the archive to include both event records and copies of the actual fax images, make the Save Pages With Event option active. To make an archive of event records only, remove the checkmark from this option.

To include copies of fax attachments with each fax event from the Send log, select the Save Attachments option. If you sent the same attachment with many faxes, it may be more efficient to save only one copy of that attachment in the archive.

To retrieve an item from an archive, select the Open button in the Send or Receive log dialog boxes. The dialog box will display the archived files in the event list. To return to the current Send or Receive log, click on the Active Log button.

In order to conserve disk space, WinFax normally compresses archived fax images and records. But since it takes longer to open a compressed file, you may decide that you don't want to compress your archive files. To disable automatic compression, add this line to the [General] line of the WINFAX.INI file:

```
Compress=0
```

Log

The event lists in the Send and Receive logs have four columns of information about each fax event. You can use the Log Display dialog box to choose the type of information in each column, and the order in which events appear in the list. Select the Display button to use the Log Display dialog box shown in Figure 13.14.

The Example box shows the current arrangement of the four column headings. To change the information in a column, drop down the list in that numbered field block and select the new heading. The default size of the new field will automatically appear in the Width field; use the up or down arrows to make the column wider or narrower. In most cases, there's no need to change the default width.

The Sort By box specifies the order in which events appear in the log. Drop down the menu to choose either Date, Destination, Time, or Status. When more than one event meets the first Sort By criterion, WinFax will sort by the elements in the second and third fields. Choose Ascending order to display the oldest events first, or to list destinations in alphabetical order. Choose Descending order to start with the latest events or to sort destinations in reverse alphabetical order.

FIGURE 13.14

The Log Display
dialog box

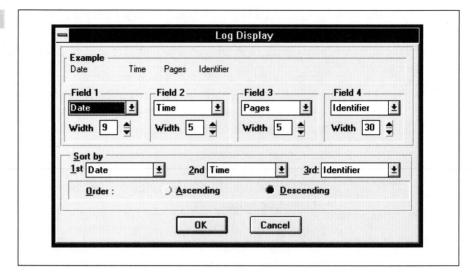

Returning to the Log box in the Send and Receive logs, you can use the Print button to create a hard copy of the current event list. The Print dialog box offers a chance to choose a printer and to specify the font size and typeface.

To store a list of records as either an ASCII text file or a dBASE file, use the Export button. When the Export Log dialog box appears, type a file name in the Export Filename field or use the Select... button to choose an existing filename. When you export in ASCII format, you can use the Field List and delimiter boxes to specify the information in each record and the separators between fields and between records.

Working with Fax Records in WinFax 4.0

WinFax 4.0 provides considerably more flexible methods for managing fax messages than the earlier versions. You can organize the information in each log to meet your own particular needs, and display the actual fax on your screen along with the transmission information, as either a thumbnail view or a readable image.

Scheduled Events

The Outbox window shown in Figure 13.15 contains a list of outgoing faxes that have not yet been transmitted. To display the Outbox Window, either click on the Outbox icon button or select Fax ➤ Outbox.

The Outbox window has two sections. The top part of the window contains a list of events in chronological order. Each event listing includes details about the scheduled event, including the scheduled time and date of transmission, the name of the recipient, the subject, and the number of pages. If you want to add or remove fields from the event list, move the cursor to the heading line and click on your right mouse button. Choose the item you want to add or remove, and add or remove a check mark. To change the order in which the headings appear, drag the button to the new position.

FIGURE 13.15

The Outbox window

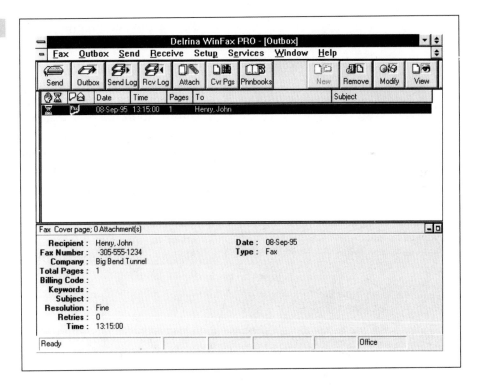

The bottom portion of the Outbox Events window may contain information about the currently highlighted event, a thumbnail view of the document, or a fully detailed view of the fax. To choose the type of display, either choose one of the Display commands in the Window menu, or move the cursor to that part of the screen and click on the right mouse button.

If you don't do anything to change it, WinFax will send each item in the Outbox list at the specified date and time. You can make the following changes to a scheduled event:

Send Now — To transmit the highlighted item immediately, select the Send Now command from the Outbox menu or from the menu that appears when you place the cursor on the highlighted event and press the right mouse button.

Reschedule — To change the scheduled transmission date or time of a highlighted event, choose Outbox ➤Reschedule.

Remove — To cancel a scheduled transmission, highlight the event listing and use either the Remove icon button or the Remove... command in the Schedule menu or in the menu that appears when you highlight an event listing and press the right mouse button.

Modify — To add, remove, or change the subject, keywords, or billing code of a scheduled event, highlight the event listing and select either the Modify... command in the Outbox menu or the Modify icon button in the toolbar.

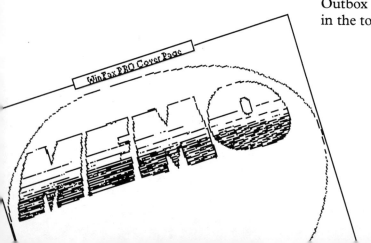

View	Either use the View command in the Fax menu or the right mouse button menu, or select the View icon button to start the Viewer program with the highlighted fax on display. If the transmission includes attached binary files, the View command will start an application that can display the contents of the file. For example, if you attach a graphic file with a .PCX extension, WinFax will start the Windows Paint program.
Hold	Either use the Hold command in the Outbox menu or the right mouse button to suspend a transmission schedule, or the Hold All command to suspend transmission of all scheduled events. When you place a Hold on an event, WinFax will not send it until you enter a Release command from the Outbox menu.
Change Destination	Use the Change Destination… command in the Outbox menu or in the right mouse button menu to send the highlighted fax to a different person or to use a different fax telephone number.
Filter	Use the Filter command in the Outbox menu or in the right mouse button menu to restrict the events in the Outbox Window to either scheduled events, held events, or both, and to include or exclude faxes for groups.

The Send and Receive Logs

After WinFax Pro 4.0 sends or receives a fax, it places a record of that event in the Send Log or the Receive Log (Figure 13.16). You can use the Send and Receive logs to retrieve copies of old faxes, display information about each fax event, including unsuccessful attempts to send or receive a fax, and send another copy of the same fax.

FIGURE 13.16

The Send/Receive Log

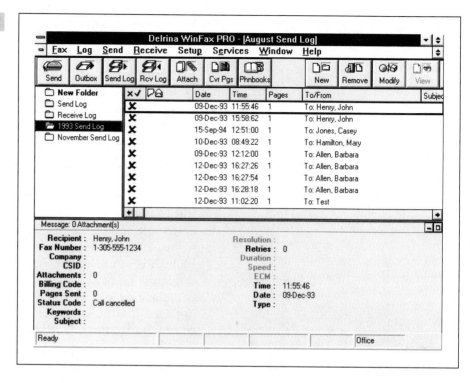

To open a log, click on the Send Log or Rcv Log icon buttons, or select the Log... command from the Send or Receive menus. To change the log currently on display, either click on the Send or Rcv Log icon buttons or select the log you want to open from the list on the right side of the window.

The Send and Receive Log windows have three sections. The list of log folders is in the upper-left part of the window. The list of events in the current folder is in the upper-right, and additional information about the currently selected event occupies the bottom part of the window.

Like the Windows File Manager, the current log appears as an open file folder. To open a different folder, highlight it and double-click with your mouse.

To create a new folder, double-click on New Folder and type a name in the New Folder dialog box. New folders may either be top-level folders or subfolders under an existing folder. To move an item into a newly created folder, highlight the listing in the item list, hold down your left mouse button, and drag it to the name of the folder where you want it to go. To keep the original where it is and place a copy in a different folder, hold down the Ctrl key while you drag and drop.

By creating new folders and moving records into them, you can keep the number of items in any folder down to a managable size. Unlike the earlier versions of WinFax, Version 4.0 automatically stores records in compressed form, so you don't need to keep archived records separate from currently active folders. You can create new folders for faxes sent or received during a period of time (such as a month), and keep the monthly folders as subfolders named for the year, or you could sort faxes by subject or recipient instead.

The rest of the top part of the Log windows contains a list of the events in the current log. A check mark in an event listing indicates that the event was successful—all pages of this fax and all attachment files were transmitted or received. An X indicates that the transmission was not complete. You can find the reason for a failure in the Status field, either in the bottom part of the Log Window or as part of the event listing.

You can configure the bottom part of the window to show either expanded information about the currently highlighted fax event, thumbnail views of the fax, or a detailed view of the fax image. To specify the type of display you want, select either the Display Information, Display Thumbnails, or Display Fax View command from the Window menu or from the menu that appears when you move the cursor to the lower part of the window and click on the right mouse button.

The meaning of most of the information fields in the event list and the expanded information display are pretty clear. The Total Pages field shows the number of pages in the document, even if WinFax was not able to successfully send or receive them. The Pages Sent field shows the number of pages that actually made it through the fax link. These two pages are equal for successful transmissions.

If WinFax receives a busy signal when it places a call, it will redial until it makes a connection. The Retries field shows the number of times WinFax attempted to make the connection.

You can use these commands to manage event records in the Send and Receive logs:

View   Choose Fax ➤ View... or click on the View icon button to start the WinFax Viewer program with the currently highlighted event in the display. You can accomplish the same thing without opening a separate window by selecting Window ➤ Display Fax View or by moving the cursor to the bottom part of the Log window and clicking on the right mouse button.

Remove   Use the Remove command in the Log menu or in the menu that appears when you click the right mouse button, or click on the Remove icon button to delete the currently highlighted fax event record or folder. When you enter a Remove command, the Delete Log Records dialog box shown in Figure 13.17 appears. When Delete Pages (Keep Event) is active, the log will retain a record of deleted fax events, but it will remove the actual images. When Delete Attachments is active, the command will remove attachment files, but it will retain the fax send files.

Modify   Use the Modify command in the Log menu or in the right mouse button menu, or click on the Modify icon button to add, change, or delete the subject, keywords, or billing code for the currently highlighted fax event.

Resubmit/Forward   Use the Resubmit/Forward command in the Log menu or in the right mouse button menu to fax a copy of the currently highlighted fax event (including attachments). When the Send Log is active, the Send Log appears with the name and fax number of the original recipient of this fax. If you want to send this fax to somebody else, edit the name and fax number fields, or choose a name from a phonebook.

If the Receive Log is active, the Send Log appears without a name or fax number already filled in. You can treat this dialog box just as if you were sending this fax for the first time.

Save   Choose Log ➤ Save Attachments... or the Save Attachments command in the right mouse button menu to save copies of fax attachment files separately from the fax record. The Save Message Attachments dialog box shown in Figure 13.18 is a Windows file selector where you can specify the name of the file and the path and directory under which you want to store the file.

FIGURE 13.17

The Delete Log
Records dialog box

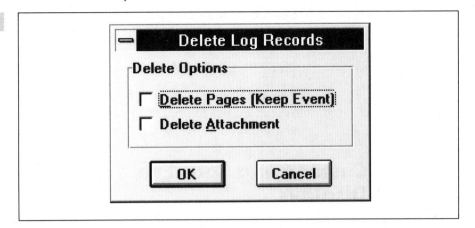

FIGURE 13.18

The Save Message
Attachments dialog
box

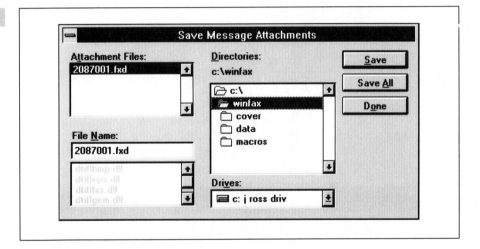

Sort   Use the Sort... command in the Log menu or in the
right mouse button menu to specify the order in which WinFax
will arrange the fax events in the logs. When more than one event
shares the same primary sort key, WinFax sorts the events by the
second and third keys.

There's an easier way to sort an event list: click on a column
heading to sort the list by that category.

Search   Use the Search command in the Log menu or in the right mouse button menu to create a log display that only includes fax events that meet specified search criteria, such as event type, recipient, subject, or date.

The 4.0 Search command functions exactly the same as the 3.0 Search button in the Display box. For details about the 4.0 Search command, see the "Display" section above.

Customizing Your Event Record Displays

You can modify almost all of the elements in the Outbox, Attachments, and Event Log windows to meet your own specific needs. Here's how to do it:

- To change the relative sizes of the folder list, event list, and display area, move your cursor to the border between any two areas, hold down the left mouse button, and drag the border to the new position.

- To quickly change the size of the display area in the bottom part of the window, click on one of the size buttons of the right edge of the window, just below the event list. Click on the button with a solid bar to minimize the size of the display area, or click on the button with an open box to maximize the display area.

- To change the width of a column in the event list, move the cursor to the right-hand border of that column, hold down the left mouse button, and drag the border to its new location.

- To change the relative positions of the columns in an event list, move your cursor to the column you want to move, hold down the left mouse button, and drag the column to its new location.

- To add or remove a column, move the cursor to the column heading and click on the right mouse button to display a menu of possible headings. When the list of possible headings in Figure 13.19 appears, the fields that are currently in use have check marks next to them; click the left mouse button to add or remove a heading.

- To display or hide the toolbar that contains the icon buttons, select Window ➤ Toolbar.

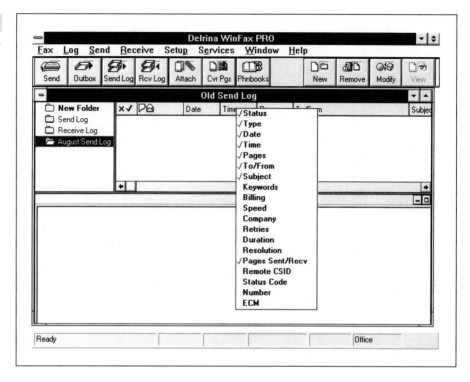

FIGURE 13.19

The Send Log column list

- To add or remove icon buttons from the toolbar, choose Setup ➤ Toolbar. You can add an unlimited number of buttons and spaces to the toolbar, but it won't do you much good, since it's not possible to display the buttons that extend beyond the right margin of the window. To add or remove the captions in the icon buttons, click on Show Toolbar Text.

- To display or hide the status bar at the bottom of the window, choose Window ➤ Status Bar.

- To change the order of the fields in the display area, click on a field and drag it to its new position.

This chapter explained WinFax attachment files and the log files that contain information about past and future fax transmissions. In the next chapter, you will learn how to more closely integrate WinFax with your Windows applications, using dynamic data exchange macros.

CHAPTER

14

Using WinFax Pro Macros

IN Chapter 11, you learned how to send faxes from Windows applications by selecting WinFax as your printer. In this chapter, we'll explain how to use the dynamic data exchange (DDE) macros included in the WinFax Pro package to integrate WinFax more closely with Word for Windows, Ami Pro, and Excel.

If you're familiar with the Windows DDE protocol, you can create your own macros to move DDE commands and data between WinFax and other Windows applications. Appendix A contains the specific DDE syntax for WinFax.

Using WinFax Macros with Microsoft Word for Windows

The WinFax Pro 3.0 package includes a macro that adds a WinFax command to one of the Word menus and a sample DDE/Merge macro that extracts recipient names and fax numbers from a merged document file. Both macros are contained in the WINFAX.DOT template file.

In WinFax Pro 4, you can install the macros by clicking on the Microsoft Word Macro icon in the Windows Program Manager. When the WORD20.DOC or WORD60.DOC template appears on your screen, click on Install Macros to add a WinFax command to Word's File menu, start the DDE/Merge macro, or both.

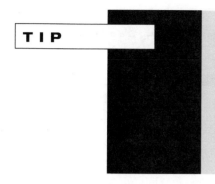

TIP

If the macros supplied with WinFax Pro don't meet your needs, you can use Microsoft's WordBasic macro language to modify the DDE/Merge macro or to write your own macros. You can obtain a free WordBasic manual from Microsoft by sending them the "Microsoft Word for Windows Supplemental Offers" card in the Word for Windows package, or by calling 800-426-9400. Ask for Kit No. 059-050-839.

In WinFax 3.0, the WIINFAX.DOT file is located in the WINFAX\MACROS subdirectory. Before you install a Word macro, you should copy this file to the WINWORD directory.

Adding WinFax to a Word for Windows Menu

The WINFAX templates include a macro that makes it possible to send faxes from Word by selecting WinFax from a menu. Follow these steps to install the macro in version 2.0 of Word for Windows:

1. Start Word for Windows.

2. Use the File menu to open the NEWMACRO.DOC document.

3. Click on the List Macros button. The list of macros shown in Figure 14.1 will appear.

4. Highlight *InstallMacro* and click on the Install button to display the dialog box in Figure 14.2.

5. Select the Normal template and click on OK.

6. Close the dialog box with the list of macros, and close the NEWMACRO.DOC document.

7. Choose Open from the File menu. The Open dialog box shown in Figure 14.3 will appear. Select Document Templates (*.dot) from the List Files of Type menu.

FIGURE 14.1

The list of Word for Windows macros

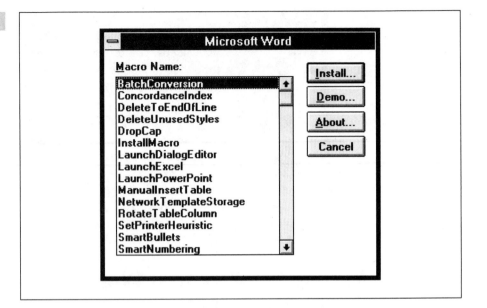

FIGURE 14.2

The Install Macro dialog box

FIGURE 14.3

The Open dialog box

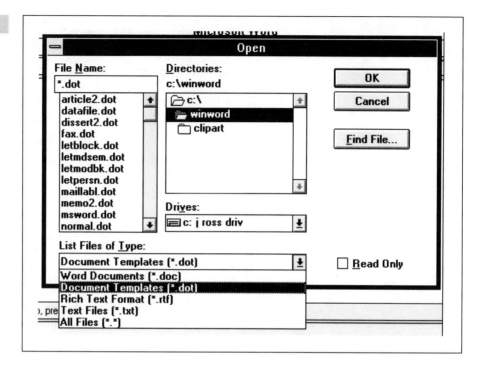

8. Select WINFAX.DOT from the list of documents and click on OK. The template appears on the screen as a blank document.

9. Choose Macro from the Tools menu. When the Macro dialog box shown in Figure 14.4 appears, choose InstallMacro and click on the Run button.

10. When the Install Macro dialog box shown in Figure 14.5 appears, choose WinFax from the Macro To Install box, highlight the Normal template and click on the Copy button.

11. Select Tools ➤ Options. The Options dialog box shown in Figure 14.6 will appear.

FIGURE 14.4

The Macro dialog box

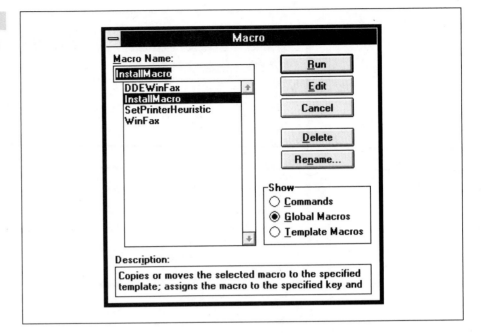

12. Click on the Menus icon in the Category list.

13. Choose WinFax from the list of macros, and choose the name of the menu where you want WinFax to appear. Since the Print command appears in the File menu, that's probably the best place to put the WinFax command.

14. If you want to create a WinFax icon button in the toolbar, select the Toolbar icon from the Category list in the dialog box in Figure 14.7. Highlight WinFax in the list of macros and choose a button icon. In the Tool To Change box, choose one of the [space] items. Finally, click on the Change button to edit the toolbar and select the Close button to close the dialog box.

This procedure installs the WinFax macro in Word for Windows 2.0. The process is different in WinWord 6.0.

After you have installed the WinFax macro, the WinFax command appears in the Word for Windows menu you specified (as shown in Figure 14.8). The WinFax command starts WinFax and displays the Send Fax dialog box. To send a fax from Word, you can now select the WinFax command instead of opening Print Setup and entering a Print command.

FIGURE 14.5

The Install Macro
dialog box

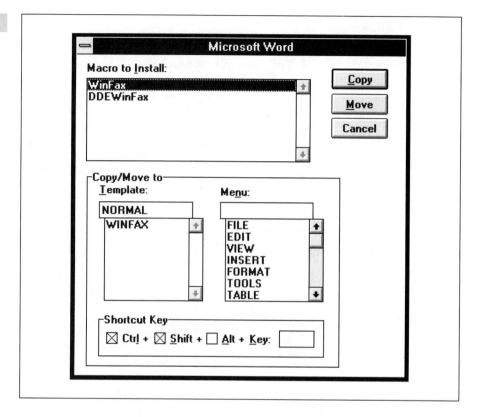

FIGURE 14.6

The Options dialog
box with menu active

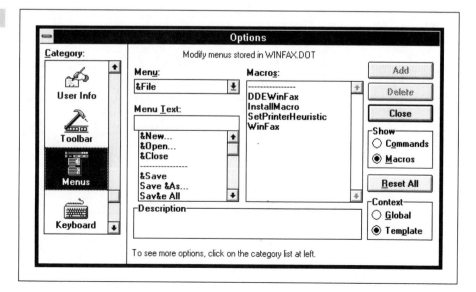

FIGURE 14.7

The Options dialog box with toolbar active

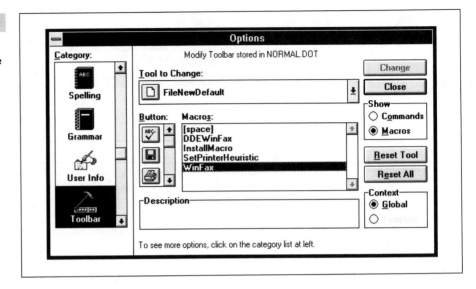

FIGURE 14.8

The File Menu with WinFax option

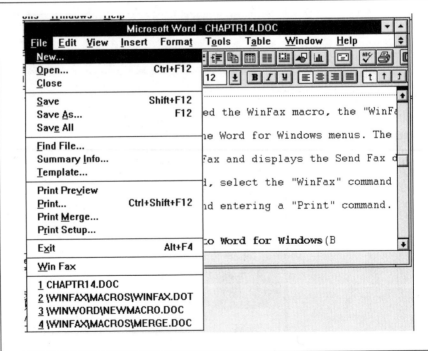

Adding a DDE/Merge Macro to Word for Windows

The Word DDE/Merge macro reads a merged document file and sends the name and fax number of each recipient to WinFax, which places the document in its pending fax events queue. To use this macro, you should be familiar with Word's Print Merge functions.

The COVERPG.DOC file is a sample fax cover sheet in Word format, which you can edit and use in place of a WinFax cover sheet. When you open the MERGE.DOC file, Word combines names and fax numbers from the MRG_LST.DOC file with the text from the COVERPG.DOC file. To extract names and fax numbers and send them to WinFax, run the DDEWinFax macro by selecting Tools ➤ Macro ➤ DDEWinFax.

If you plan to modify the DDEWinFax macro, copy it to the Normal template first by choosing Macro from the Word Tools menu and running InstallMacro. Make your changes to the NORMAL.DOT file rather than to the original WINFAX.DOT file.

Adding WinFax to an Excel Menu

The WINFAX2.XLM file adds two commands to the Excel 4.0 File menu. The WNFX1XL3.XLM and WNFX2XL3.XLM files each add one command to the Excel 3.0 file menu.

Follow these steps to install an Excel macro:

1. Copy the macro file or files for your version of Excel from the WINFAX\MACROS subdirectory to the \EXCEL\XLSTART subdirectory.

2. Start Excel. The WinFax commands will appear in the Excel File menu, as shown in Figure 14.9.

You can use the WinFax Setup command (which displays the dialog box Figure 14.10) to specify the port connected to your fax modem and to add a telephone icon button to the Excel button bar.

FIGURE 14.9

The Excel File Menu with the WinFax options

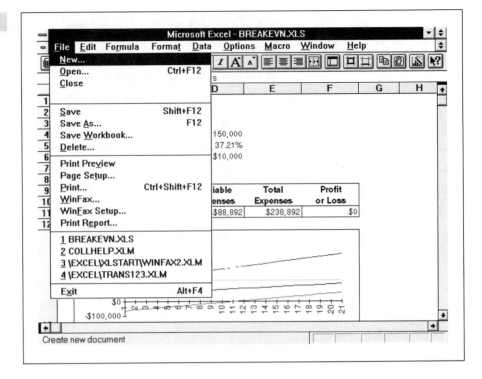

FIGURE 14.10

The WinFax Print Setup dialog box in Excel

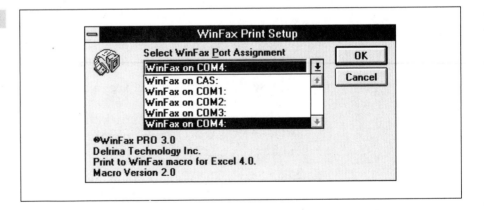

The WinFax command instructs Excel to send the current worksheet to WinFax. When you choose WinFax from the File menu or (if you made it active in WinFax Setup) click on the telephone icon button, Excel displays its Print dialog box with WinFax as the default printer. When you click on OK, the Send Fax dialog box appears.

Using WinFax Macros with Ami Pro

WinFax Pro includes two Ami Pro macros. The WINFAX macro adds a command that sends the current document directly to WinFax. The DDEWNFAX macro is a sample DDE/Merge command.

In WinFax 3, complete this procedure to install the WINFAX macro. (In WinFax 4, click on the Ami Pro icon in the WinFax group of the Windows Program Manager, and start with step 4. If Ami Pro doesn't start, use File ➤ Properties in Windows Program Manager to add the AMIPRO directory to the command line.)

1. Copy the WINFAX.SMM file from the WINFAX\MACROS subdirectory to the AMIPRO\MACROS subdirectory.

2. Copy the WINFAX.BMP file from the WINFAX\MACROS subdirectory to the AMIPRO\ICONS subdirectory.

3. Start Ami Pro.

4. Choose the SmartIcons command from the Tools menu. The SmartIcons dialog box shown in Figure 14.11 will appear. When the display of .BMP files appears, drag the WinFax button (the last one in the Available icons box) to the Default box.

5. Select Edit Icon to open the Edit Icon dialog box. Choose WINFAX.SMM from the list of macros, and click on OK. Click on OK again to get out of the SmartIcons dialog box.

6. Run the AMIMENUS macro from the Playback command in the Macros subdirectory under the Tools menu. When the Customize Ami Pro dialog box shown in Figure 14.12 appears, highlight *&File* in the Menus box and click on Add An Ami Pro Function Or Macro....

FIGURE 14.11

The SmartIcons dialog box

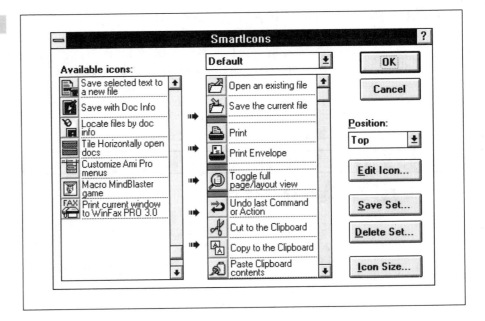

FIGURE 14.12

The Customize Ami Pro dialog box

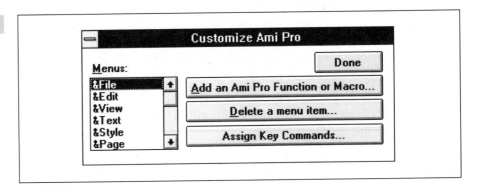

7. When the Add Menu Item dialog box shown in Figure 14.13 appears, type **&WinFax** in the Name For Menu field. Use the Insert Before field to specify the location where you want the WinFax command to appear in the menu.

8. Click on OK to close all open dialog boxes.

After you have installed the WinFax macro, you can send faxes from Ami Pro by selecting File ➤ WinFax or by clicking on the Fax icon button in the button bar.

FIGURE 14.13

The Add menu Item
dialog box

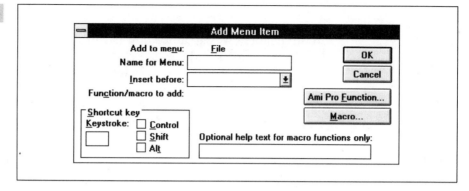

Follow these steps to use the DDE/Merge macro:

1. Copy either the DDEWNFAX.SM file (for Ami Pro 3.0) or the MDDEWNFX2.SM file (for Ami Pro 2.0) from the WINFAX\ MACROS subdirectory to the AMIPRO\MACROS subdirectory.

2. Start Ami Pro.

3. Choose the SmartIcons command from the Tools menu.

4. Select Edit Icon to open the Edit Icon dialog box. Choose WIN-FAX.SMM from the list of macros and click on OK.

5. If you want to add the DDE/Merge function to a menu, use the AMIMENUS macro, just as you did for the WINFAX macro.

Using WinFax with Lotus 1-2-3

The WinFax package doesn't include any Lotus 1-2-3 macros, but it's easy to create one of your own. To automatically fax a document created in 1-2-3 for Windows, mark the print area and make WinFax your default printer, and use this macro:

```
{DDE-OPEN "FAXMNG","TRANSMIT"}
{DDE-POKE recipient,SENDFAX}
{Alt}fp~
```

In the second line, use the name of the cell that contains the recipient's fax number in place of *recipient*.

In this chapter, we explained how to install the macros supplied with Win-Fax in Word for Windows, Excel, and Ami Pro. When the macros are in place, you can send a fax by selecting a command from a menu or by clicking on a toolbar icon button. If you're familiar with the Windows Dynamic Data Exchange protocol, you can write your own macros to integrate WinFax with other Windows applications, using the syntax in Appendix A.

CHAPTER

15

Advanced WinFax Pro Functions

WHEN you receive a fax, you probably read it (or at least skim the text) as soon as it arrives to decide whether you want to save and/or reply to it. If you receive the message on a conventional fax machine, you might write your reply directly on the received fax message and either fax it back to the originator or send it along to somebody else. WinFax includes two special functions that make it possible to do the same thing with your fax modem. You can use the WinFax annotation feature to add text and drawings to a WinFax fax image. WinFax also supports scanners compatible with the TWAIN standard, so you can write notes on a paper copy of a received fax, scan the marked-up copy, and resend it. You can use the same scanner support to fax copies of letters, newspaper clippings, and other documents that did not originate in your computer.

Sometimes you receive faxes containing text that you want to include in a document of your own, either as received or with editorial changes. WinFax includes an optical character recognition (OCR) feature that converts text in a fax image to machine-readable text that you can export to a word processor or desktop publisher.

In this chapter, we'll explain how to annotate fax images, how to use a scanner with WinFax, and how to use OCR to convert text. However, it's important to remember that WinFax isn't always the most convenient way to handle these faxes. If you have a stand-alone fax machine, it may be faster and easier to send a paper document through it than to scan the document into WinFax. And if you're sending text to somebody else who has a fax modem, it will be considerably faster to send it as an ASCII or binary file using your data communications program or the WinFax Pro 4.0 binary file transfer features.

Annotating Fax Messages

When you annotate a fax image, you add text, graphic images, or freehand drawings to the existing image. You can either make the annotations a permanent part of the original image, or you can save the annotations as an overlay that doesn't change the image itself.

Remember Luigi's Pizzas from Chapter 1? If you don't have one of his menus, he'll send you one by fax. He wants you to check the ingredients you want on that special combination, fill in your name, address, and telephone number, and fax it back to him. If you're using WinFax, you'll need to use the WinFax Annotation feature to mark up one of Luigi's menus. Fortunately, Luigi hasn't figured out how to deliver your pizza by fax, so you won't need to worry about cleaning the tomato sauce out of your modem.

When you receive the menu by fax, it appears in the fax event list shown in Figure 15.1. To annotate it, highlight the event listing and start the Viewer program.

Select Annotate ➤ Show. This displays the annotation toolbar, adds an annotation icon to the existing button bar, and changes the commands in several Viewer menus. You can choose tools from either the toolbar or the menus.

FIGURE 15.1

The WinFax Pro 3.0 Adminstrator screen with one item in the event list

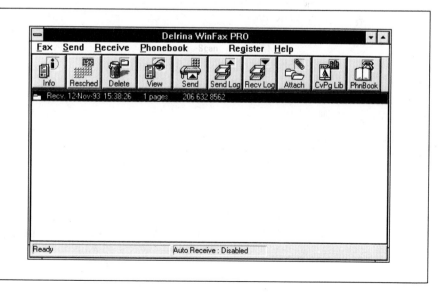

Adding Text to an Image

To add text to a fax, first use the text tool to create a text object. Select the text tool icon, and then move your mouse to the location where you want the text to appear. Hold down the left mouse button and drag the mouse to define the edges of the text object box.

When you create a new text object, a flashing cursor appears inside the box. Depending on the text alignment, the cursor is at either the left, center, or right of the box. To change the cursor's starting location, choose a different alignment icon button (shown in Figure 15.2), or select Left Justify, Center Justify, or Right Justify from the Text submenu under the Annotate menu.

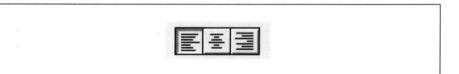

To add text to or edit an existing text object, click on the text editing tool icon button and move the cursor to the location where you want to add or change the text.

To change the font, highlight the text and choose a new font name from the font selector in the toolbar. The number to the right of the typeface name specifies the type size. Use the B button to boldface selected text or the I button to italicize it. The U button underlines selected text. Figure 15.3 shows the type style icon buttons.

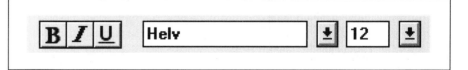

You can tell Luigi where to deliver your pizza by creating a text object over the name and address section of the order form and typing your information in the appropriate lines, as shown in Figure 15.4.

FIGURE 15.4

A text object on Luigi's pizza menu

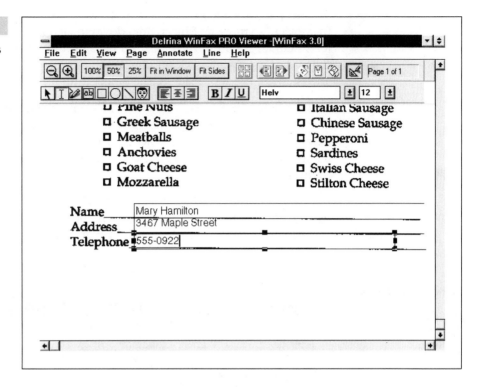

Drawing on an Image

Sometimes it's easier to draw a picture than to describe something with text. For example, you might want to order half a pizza with anchovies and half without. You can use the pencil tool and graphic tools to draw a picture that tells Luigi just exactly how he should arrange the toppings on your pie.

When you add a graphic to a fax image, you create an annotation object. To add an object, select the icon button with the type of graphic you want to add and move your mouse to one corner of the location where you want to place the object. Then hold down the left mouse button and drag the mouse to the opposite corner. The outline of the new object will appear on your screen.

To select an existing object, click on the Select icon tool button and click on the object. To move an object, hold down the left mouse button while the selector cursor is within the object and drag it to its new location. To change the size of an object, click on one of the handles on the edge of the object and drag the corner or side of the object until it is the desired size.

You can select more than one object at a time by clicking on the first object and then holding down the Shift key when you select additional objects. To delete a selected object or objects, use the Delete command in the Edit Menu. To place an object in the Windows Clipboard, use the Cut command in the Edit menu.

For example, you could place a circle on a blank space in the image and then use the pencil tool to show anchovies on one section with green peppers and mushrooms on the other, as shown in Figure 15.5.

FIGURE 15.5

The Pizza menu with the marked-up pizza drawing

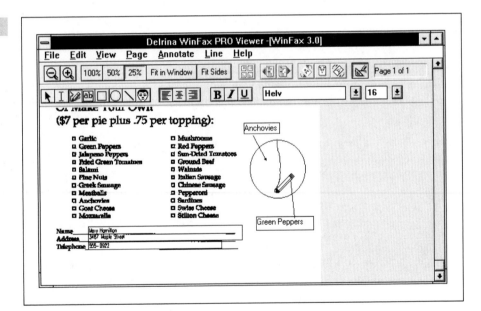

To add check marks in the boxes next to the toppings you want to order, click on the pencil tool and make an X in each box, as shown in Figure 15.6. If the lines aren't thick enough, you can use the line weight command in the line menu to change them.

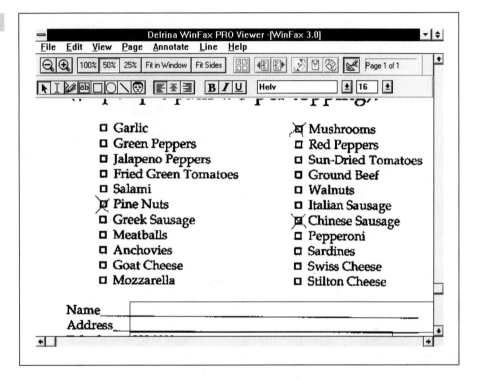

You can import a graphic from a draw or paint program by selecting the graphic tool icon button shown in Figure 15.7 and choosing the file you want to import.

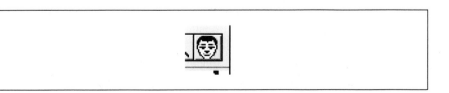

Annotation Commands

If you prefer, you can select annotation commands from the Viewer menus, instead of using the toolbar. There are also some commands in menus that don't appear on the toolbar. These commands appear in various menus when you select the Show command in the Annotate menu.

Here's a summary of annotation commands.

File Menu

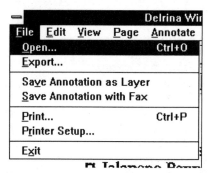

Save Annotation As Layer (WinFax Pro 3.0 only)   When you save your annotations as a layer, rather than as part of the original fax image, it's still possible to print or re-send the original fax image without the annotations. After you have created your annotations, click on this command to save them separately from the image.

To view the fax image without the annotation layer, select the Hide command in the Annotate menu, or click on the Annotate icon.

Save Annotation With Fax (WinFax Pro 3.0 only)   This command is the opposite of Save Annotation As Layer. When you use it, the annotations become a permanent part of the original fax image.

Edit Menu

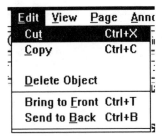

Bring To Front   When several objects overlap, the Bring To Front command moves the currently selected object to the top.

Send To Back   When several objects overlap, the Send To Back command moves the currently selected object to the bottom.

View Menu

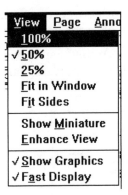

Show Graphics   When Show Graphics is active, annotation graphics are visible on your screen and in printed copies of the annotated image. If you're working on a complicated annotation, you might want to hide graphics to allow WinFax to redraw the image more quickly.

Fast Display   When Fast Display is active, graphic annotations appear more quickly, but less accurately, than when Fast Display is not active.

Annotate Menu

Show   The Show command turns annotation on. When you enter the Show command, WinFax adds all of the other annotation commands to the Viewer menus and displays the annotation layer over the current image.

Hide   The Hide command turns off annotation, removes the annotation layer, and makes the annotation commands disappear.

Shade   The Shade command specifies the content of the current object and all additional objects until you enter another Shade command.

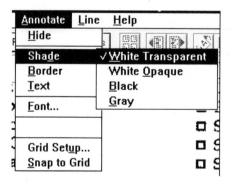

White Transparent objects are outlines. When they overlap other objects, you can see the second object through the transparent object.

White Opaque objects are solid white. Other objects are not visible through them.

Black objects are solid black. Gray objects are solid gray. Other objects are not visible through black or gray objects.

Border   The Border command specifies the type of black line around the edges of an object.

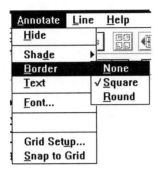

Text   The Text submenu specifies the color, alignment, and style of the currently selected text. Except for White Text and Black Text, the commands in the Text submenu are the same as the ones in the Annotation tool bar.

The White Text command reverses the currently selected text, making the background black and the text white. This can be used to attract attention to an annotation by creating a contrast with the text in the original fax image.

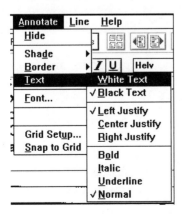

Font... The Font Command starts the Font dialog box, from which you can specify the name, style, and size of the currently selected text. It's probably easier to use the tool bar to specify the font.

Spelling... The Spelling command starts the spell checker. To confirm that you have spelled everything in your annotation correctly, highlight the text and select the Spelling...command. WinFax will identify misspelled words and suggest possible corrections.

Graphic... Use the Graphic command to import an existing graphic file into the currently selected graphic object, which you created with the Graphic icon. When you enter the Graphic command, the Graphic Attributes dialog box shown in Figure 15.8 appears, which lets you specify the source of the graphic file, and the way it will appear as an annotation. You can also start the Graphic Attributes dialog box by double-clicking on an object.

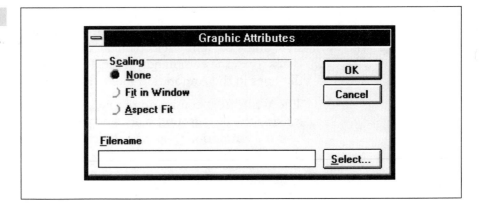

Stamps The Stamps command inserts a bitmap graphic image (which can be a "rubber stamp" annotation such as an "approved" mark or a signature).

Grid Setup... The grid is a set of invisible vertical and horizontal lines that force the exact location of graphic objects. Use the Grid Setup command to set the amount of space between grid lines. When Show Grid is active in the Grid Setup dialog box, the grid is visible on your screen, but it does not print.

Snap To Grid   When the Snap To Grid command is active, objects move to the closest grid lines when you create or move them.

Line Menu

The Line menu (shown in Figure 15.9) shows three options for line thickness and four for line ends. When you choose a new thickness or line ending, it changes the currently selected object and all new objects until you select a different weight or ending.

The line thickness options change the weight of straight and curved lines and the borders of boxes, text objects, ovals, and graphic objects. Line endings only apply to straight lines.

FIGURE 15.9

The Line Menu

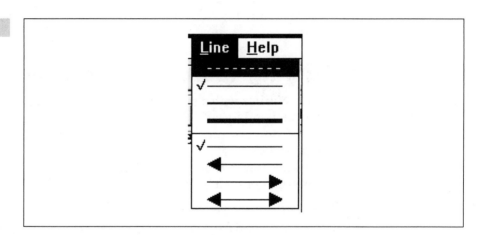

Using a Scanner with WinFax

Not everything that you want to send by fax originates as a computer document. In order to use WinFax to fax from paper copies (or for that matter, cloth, plastic, or other material), you can use a TWAIN-compatible scanner to scan the document into your computer. It's also possible to use the OCR (recognize) function to export the text of a scanned-in document to other applications as ASCII text.

Once again, let's use Luigi's pizza menu as an example. Instead of receiving it by fax, you might pick up one of his order forms at the restaurant. When you're ready to order, you check off the toppings you want with a pencil and feed the marked-up menu into your scanner, which transfers an image to WinFax. Or if you prefer, you could scan in the menu without marking it up, and use annotation commands (explained in the previous section of this chapter) to tell Luigi that you want him to put sardines, fried green tomatoes, and Swiss cheese on your pizza.

In order to use a scanner with WinFax, the scanner driver must use the TWAIN standard to communicate with Windows. TWAIN (the word is supposed to indicate that the standard brings together **TW**o entities— **A**pplications and **IN**puts) is a standard architecture for connecting applications and devices. Scanners made by Logitech, Hewlett Packard, and Microtek, among others, have TWAIN drivers that are compatible with WinFax.

Configuring Your Scanner for WinFax

To use WinFax with a scanner, make sure the scanner has been properly connected and installed, and (if it has one) turn on the scanner's power switch. Configure the scanner for black and white, one bit per pixel image, with a resolution of 200×200 dots per inch. Set the image type as black and white drawing, line drawing, or line art. Use the scanner's software to save this configuration as the WinFax default.

If more than one scanner is connected to your computer, select the Select... command from the WinFax Pro 3.0 Scan menu or the Select Scanner... command from the WinFax Pro 4.0 Scan menu to choose the scanner you want to use. (If you don't have a TWAIN scanner, you won't see the Scan menu.)

Scanning an Image to WinFax

Once the scanner is configured, you can send a scanned image as a fax, or you can save it as a WinFax attachment. The Scan And Send... and Scan And File... commands are in the WinFax Scan menu.

When you enter one of the Scan commands, the scanning software's main window appears. After you use the scanner software to specify the image's appearance, click on the button that instructs the scanner software to perform a final scan. When the scan is complete, either the Send Fax dialog box or the Save To File dialog box will appear. Complete the dialog box just as you would for a document that originates in your computer.

Converting a WinFax Image to Editable Text

Regardless of their content, WinFax treats fax images as graphics. Your eye converts many of those marks as words, but as far as the computer is concerned, they're just a pattern of light and dark areas. As received, you can't expect to edit a fax with your word processor.

The WinFax optical character recognition (OCR) feature converts text in fax images to a form that you can export to an editor, word processor, or other application. It also includes a simple text editor of its own, so you can edit a document before you save it as a text file.

OCR is not limited to received faxes. In the previous section of this chapter, we explained how to scan documents into WinFax. After you save a scanned document as a fax image, you can use the Recognize commands to convert the text to editable form.

While OCR can save you a lot of time when you want to add somebody else's text to one of your documents, it's not always the best way to do it. If the person sending you the document also has a modem and the original document was created with a word processor, desktop publisher, or other PC application, it might be faster and easier to use your data communications program to transfer the document as a text file or a binary data file. Since fax images are limited to 200 dpi resolution, you'll get a better printout from a data transfer, especially if you have a laser printer.

Follow these steps to convert text from a WinFax image:

1. Display the fax image you want to convert in the WinFax Viewer.

2. Select Recognize ➤ Setup.... The Setup dialog box box shown in Figure 15.10 will appear.

FIGURE 15.10

The Setup dialog box

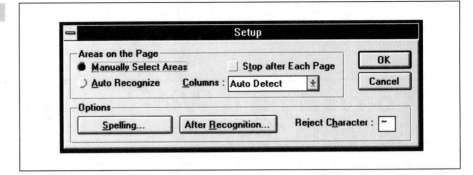

3. In the Areas On The Page box, choose Manually Select Areas to convert only a section of the page, or Auto Recognize to convert all the text on the page.

- Choose Manually Select Areas to convert text from part of a page. If Stop After Each Page (Apply Selected Areas To All Pages in WinFax Pro) is not active, WinFax will recognize text from the same segment of every page until you specify a different area.

- Choose Auto Recognize to convert all text on the page. If you choose Auto Recognize in Version 3.0, drop down the list of Columns options:

- No Parsing treats the entire page as one block of text.

- Auto Detect automatically identifies columns, but displays the text in a single column.

- Single/Tables inserts tabs between columns in a table.

- The Mult Col options recognize up to the specified number of columns and displays the text in that number of columns.

4. Click on the Spelling button to display the dialog box in Figure 15.11. Specify the spell checker dictionary and select the spell check options that you want to make active during recognition.

The USENG.MD dictionary uses American spelling; UKENG.MD uses British spelling. If you're in Canada or Australia, you're on your own.

FIGURE 15.11

The Spell dialog box

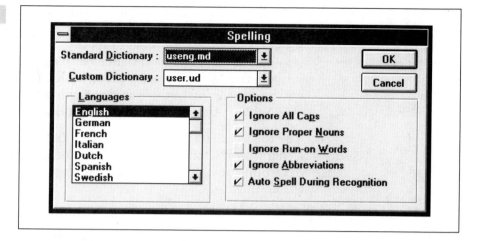

5. Click on the After Recognition button in the Version 3.0 Recognition Setup dialog box. When the dialog box in Figure 15.12 appears, specify what you want WinFax to do with a converted text. There are similar options in the 4.0 Recognition Setup dialog box.

- Choose Interactive Text Edit to display the converted text on the bottom of the Viewer screen. You can edit the converted text before you save it.

- Choose Place Text On Clipboard to use the Windows Clipboard. You can paste text from the Clipboard to another Windows application.

FIGURE 15.12

The After Recognition dialog box

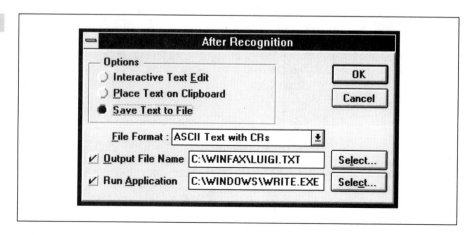

- Choose Save Text To File to store the converted text in a text file in the format specified in the File Format field. Use the Output File Name field to specify the destination of the converted text file. If you want to display the converted text in another Windows application, such as an editor or word processor, make the Run Application field active and use the Select... button to specify the application.

6. The Reject Character is the character that WinFax uses in place of a character it is unable to recognize. For example, if the program has trouble reading Luigi's menu, the word "Garlic" might appear as "Ga~lic".

7. When you're satisfied with the recognition setup, click OK.

8. To convert the text on the page currently in the display, select Recognize ➤ Current Page. To convert several pages, use the Select Pages command and specify the pages you want to convert.

9. If Manually Select is active in the Setup dialog box, WinFax will display the first page you want to convert along with a toolbar, as shown in Figure 15.13. Use the tools in the toolbar to mark the areas of the page that contain text that you want to convert.

   To change the order in which converted text will be organized, click on the Order Zones tool (the one with the # symbol) and then select the zones in the order you want to convert them.

FIGURE 15.13

The Zone selection screen

If the page display is upside down, use the Flip Page tool (the one with the two arrows) to turn it around.

You can designate each zone as either Alphanumeric or Graphic. If you specify one or more graphic zones, the Save Graphic Zones As... dialog box shown in Figure 15.14 will appear when you click on the Recognize button.

When you have specified the areas you want to convert, click on the Recognize button.

10. When you click on Recognize, or if Auto Recognize is active, the OCR process will begin.

FIGURE 15.14

The Save Graphic dialog box

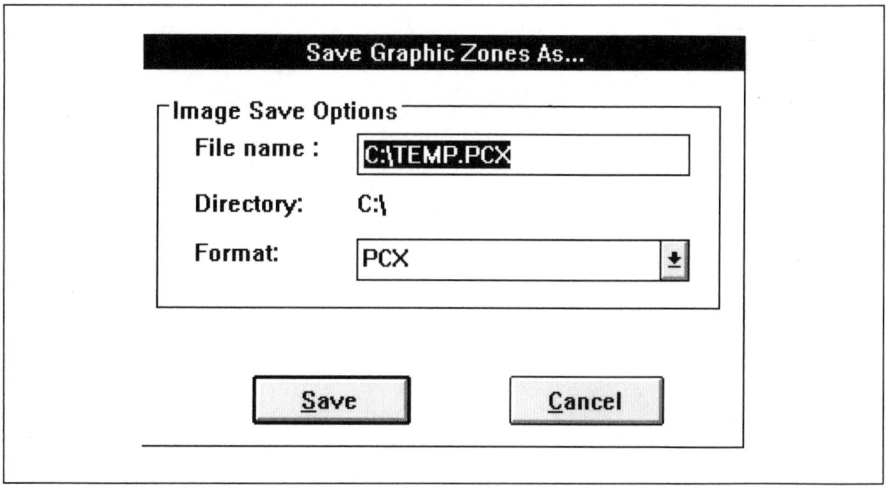

Annotation, scanning, and recognition are all advanced techniques for working with WinFax image files that increase the program's flexibility. You can add your own comments and additions to a received fax, use a scanner to import text or images, and convert received fax images to text. None of these are essential to sending and receiving faxes, but taken together, they allow you to do everything with WinFax that you can do with a stand-alone fax machine.

In the next chapter, we'll talk about the new features that Delrina has added to Version 4.0 of WinFax Pro.

CHAPTER

16

New Features in
Winfax Pro 4.0

VERSION 4.0 of WinFax Pro, released early in 1994, includes a major redesign of the user interface and a basket of new features. In this chapter, we'll explain how to move around the new screens and how to use the new features that were introduced in Version 4.0.

In addition to all the functions that were available in Version 3.0, the newest edition of WinFax Pro has these new features:

- New Screen Layout   The WinFax screens now resemble the traditional Windows desktop, where you can display several activities at the same time. The WinFax Pro screen is very similar to the screens used by other popular Windows applications.

- Right Mouse Button Menus   When you click on your right mouse button, WinFax displays a location-sensitive menu.

- Quick Cover Page   In addition to the library of graphic cover pages, you can select a simple text-only cover page.

- Improved File Attachment Files   It's no longer necessary to convert files to attachment format before adding them to a fax.

- Multiple Dial Prefixes   When you use WinFax on a portable computer, you can pre-program separate dialing codes for different telephone lines at different locations.

- Customized Credit Card Configuration   Many long-distance carriers require unconventional dialing sequences. You can use the Dial Credit Card Setup dialog box to define the specific sequence your carrier needs.

- Binary File Transfer   In addition to fax images, WinFax can also send and receive copies of data files through your fax modem in binary form, using Binary File Transfer (BFT), Microsoft At Work, or proprietary WinFax protocols.

- Polling   WinFax can connect to another PC running WinFax and instruct the distant system to send you copies of faxes that it has received.

- Automatic Forwarding   When WinFax receives faxes, it can automatically resend them to another fax number.

- Error Correction Mode   WinFax automatically eliminates transmission errors caused by noisy telephone lines or other interference when both the originator and receiver of the fax are using Class 1 fax modems. Since you don't need to do anything to take advantage of this feature, there are no commands or messages related to its use.

- "Rubber Stamp" Graphics   As part of the Viewer program's Annotation function, you can create a set of frequently used graphic images, such as a scanned-in signature or a logo, and add them to a fax with just a few keystrokes or mouse clicks.

In the rest of this chapter, we'll give you detailed instructions for using each of these features.

Moving around in the WinFax 4.0 Screen

The WinFax Pro 4.0 screen looks a lot more like the Windows desktop and other Windows application programs than it did in earlier versions of the product. Delrina calls the new screen a "multiple document interface," which makes it sound as if you can display more than one fax at the same time, which is not always true. Regardless of the name, the basic design will be familiar if you've been using other Windows applications.

Like the Windows desktop, the WinFax screen is a background that can hold several programs at the same time. Commands in the Window menu control the way WinFax arranges the Outbox, Attachments, Cover Pages, Phonebooks, and Log windows.

Changing the Toolbar

When the Toolbar command in the Window menu is active, a toolbar with up to twelve icon buttons is visible at the top of the desktop. Clicking on an icon button has the same effect as entering a command from one of the menus.

You can customize the toolbar by adding other commands that you use frequently or by removing icon buttons that you don't use very often. The Toolbar... command in the Setup menu opens the Toolbar Setup dialog box shown in Figure 16.1, which specifies the buttons used in the button bar.

FIGURE 16.1

Toolbar Setup dialog box

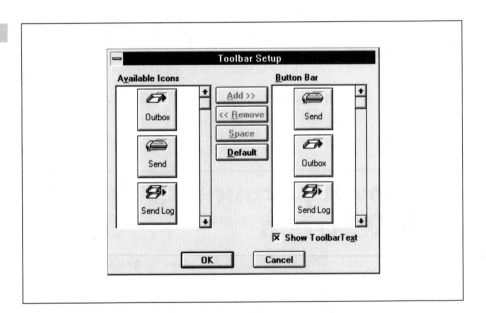

The icons in the Button Bar field are in the same order as they appear in the toolbar. To add a button to the toolbar, click on that icon in the Available Buttons field, and click on the Add>> button. To add a space, click on the Space button. If you don't specify a location for the new button or space, it will appear at the bottom of the Button Bar field, which means it will be at the extreme right of the button bar. To place a button somewhere else on the toolbar, either drag the button from the Available Buttons field to the position where you want it to appear in the Button Bar field or highlight the button in the Button Bar field that you want to appear to the right of the new button. You can add an unlimited number of buttons and spaces to the toolbar, but only the first twelve will be visible on a full-screen display.

To return to the "standard" button bar layout, click on the Default button. Click on OK to transfer your changes from the dialog box to the toolbar.

Reading the Status Bar

The optional status bar at the bottom of the screen contains information about the current status of faxes that WinFax may be sending or receiving. The Auto Receive field in the center of the status bar shows whether WinFax will automatically answer the telephone line when it rings.

To hide the status bar, click on the Status Bar command in the Window menu to remove the check mark.

Viewing WinFax Windows

WinFax can display as many as five different windows at the same time. To display a window, select a command from the menu or click on an icon button, as shown in Table 16.1.

TABLE 16.1: Commands for displaying WinFax windows

TO OPEN THIS WINDOW:	SELECT THIS COMMAND:	OR CLICK ON THIS ICON BUTTON:
Attachments	Fax ➤ Attachments...	Attach
Scheduled Events	Fax ➤ Scheduled Events...	Sched.
Cover Pages	Fax ➤ Cover Pages...	Cvr Pgs
Phonebooks	Fax ➤ Phonebooks...	Phnbooks
Send Log	Send ➤ Log...	Send Log
Receive Log	Receive ➤ Log...	Rcv Log

You can arrange windows on the WinFax screen the same way you arrange groups in the Windows Program Manager. Use commands in the Window menu to tile or cascade windows and to arrange the icons for minimized windows. The title bars on the WinFax windows follow all the standard Windows conventions—click on the triangle buttons in the upper-right corner to maximize and minimize windows; open the control menu with the bar in the upper-left corner to display control commands; and so forth.

Configuring the Item Lists

The upper portion of each window in the WinFax desktop includes a list of names or fax events called an *item list.* You can select the information fields in these item lists by moving your cursor to the column headings and clicking on your right mouse button. When the menu of fields appears, click the item you want as a heading for this column with your left mouse button.

To change the order of the columns, click on a column heading and drag it to the new position where you want to place it.

To change the width of a column, move the cursor to the border between the column you want to change and the next column to the right. When the shape of the cursor changes, drag it to the new edge of the column.

To sort an item list according to the criteria in a column heading, double-click on that column heading. For example, to sort the Send Log by the names of recipients, double-click on the To/From column heading.

You can re-send any item in a list by highlighting it and dragging it to the Send Icon button.

Configuring the Folder List

All of the WinFax windows except the Outbox have Folder Lists in the upper-left portion of the window. Folder lists are directories of folders that contain the type of event described in the same window. You can change the relative sizes of the folder list and the item list by moving your cursor to the border between the two areas and dragging it to the new position.

If one or more folders in the folder list contains sub-folders as well as events, you can display the list of sub-folders by highlighting a folder and selecting the Expand Folder command in the second menu from the left (this menu has a different name for each active window). If the sub-folder is currently visible, you can hide it with the Collapse Folder command.

To create a new folder or sub-folder, double-click on New Folder at the top of the list of folders. When the New Archive Folder dialog box appears, select the Subfolder Of option and type a description of the new folder in the Folder Name field. Press the Tab key to create a file name for this folder.

After you create a new folder, you can move one or more items into it by highlighting them in the item list and dragging the description to the folder name in the folder list.

Configuring the Display Area

The bottom part of the WinFax window contains information about the currently highlighted item in the item list. In most windows, the display area may contain thumbnail views, a full view of the current item, or a display of information about the current item. You can change the type of display by choosing a command in the Window menu, or you can make the change from the menu that appears when you move the cursor to the display area and click on the right mouse button.

When you choose the Display Fax View command, WinFax starts a program that runs the currently highlighted fax event and places the display from that program in the display area. For example, if the currently highlighted item is a fax image, the display area contains a smaller version of the WinFax Viewer program. If it's an attached graphic file, the display area contains a small version of a graphic program such as Windows Paint.

To change the relative sizes of the display area and the item list, click on one of the small buttons in the upper-right corner of the display area, or move the cursor to the top border of the display area. When the shape of the cursor changes, drag the border to its new position.

Using Right Mouse Button Menus

In earlier versions of WinFax, the right mouse button did nothing. But in WinFax Pro 4.0, the right mouse button opens a location-sensitive menu (such as the one in Figure 16.2) that contains a list of commands that apply to the current location of the cursor. To select a command from the menu, click on the command with the left mouse button. To make the menu disappear without selecting a command, move the cursor out of the menu and click on the left mouse button.

FIGURE 16.2

A typical Right Mouse Button menu

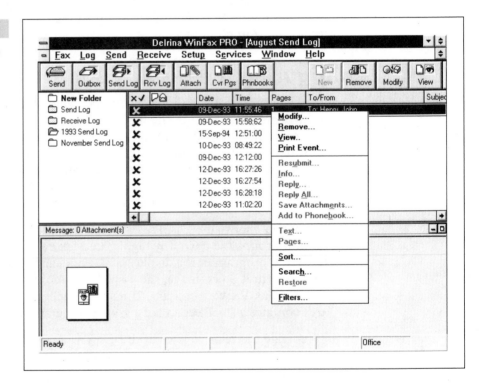

When the cursor is in a location where a menu is not available, clicking the right mouse button may display a short explanation of the field where the cursor is located.

Using Quick Cover Page

The Quick Cover Page is a simple cover sheet that is an alternative to the graphic cover sheets in the cover page library. You can customize the Quick Cover Page by adding a logo or a block of text to the plain vanilla document. Figure 16.3 shows the Quick Cover Page without any customization.

FIGURE 16.3

The Generic Quick Cover Page

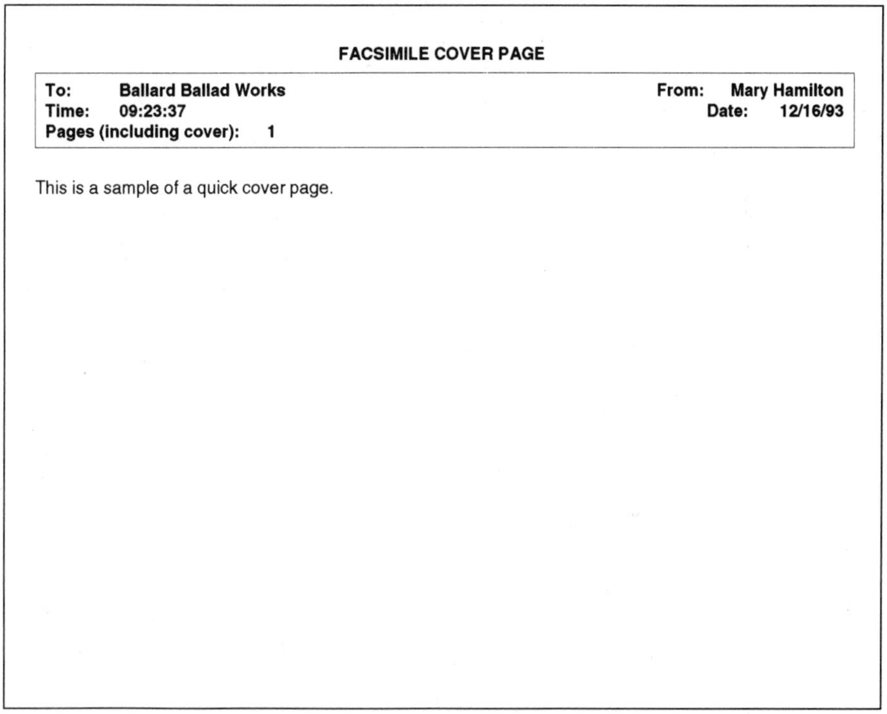

FACSIMILE COVER PAGE

To:	**Ballard Ballad Works**	**From:**	**Mary Hamilton**
Time:	09:23:37	**Date:**	12/16/93
Pages (including cover):	1		

This is a sample of a quick cover page.

To set up your own custom Quick Cover Page, open the Cover Pages window and double-click on Quick Cover Page in the item list. The Configure Quick Cover Page dialog box shown in Figure 16.4 will appear.

The bottom half of the Configure Quick Cover Page dialog box specifies the text that appears on the cover sheet. Either choose the Load From File option and click the Select... button to use text from an existing file, or

FIGURE 16.4

The Configure Quick
Cover Page dialog box

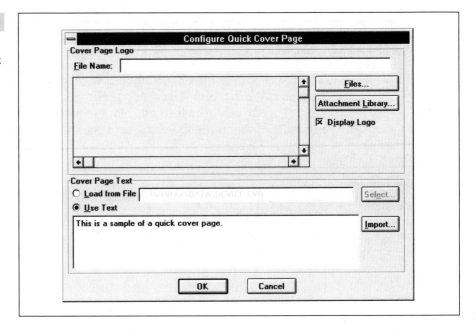

choose the Use Text option and type your text in the box under this option. The Import... button opens the same file selector as the Select button did when Load From File was active, but you can edit the imported text for use as a fax cover sheet without changing the original file. You will be able to edit this text again from the Send Fax dialog box when you prepare to send a fax.

Graphic images in Quick Cover Pages must be in WinFax Attachment format rather than in their original graphic file formats. To convert a graphic to attachment format and attach the image to a Quick Cover Page, follow these steps:

1. Start a Windows application program that can read the file. This might be the application that created the graphic, such as Corel-DRAW, or a general-purpose graphics program such as Windows Paintbrush.

2. Open the file that contains the graphic image.

3. Use the Printer Setup command (probably in the File menu) to select WinFax as your current printer.

4. Select the Print command (again, usually in the File menu).

5. When the WinFax Send Fax dialog box appears, click on the Make Attachmnt button. WinFax will create an attachment file that contains the graphic image.

6. Start WinFax and open the Cover Pages window.

7. Double-click on Quick Cover Page in the item list. When the Configure Quick Cover Page dialog box appears, either type the name of the attachment file that contains the logo in the File Name field, or click on Files to find it in a file selector. Click on Display Logo to see the image in the dialog box.

8. Click on OK to save the image as part of the Quick Cover Page.

To use the Quick Cover Page as your cover sheet when you send a fax, make Cover Page active in the Send Fax dialog box, click on the Cover... button to display the Select Cover Page dialog box shown in Figure 16.5, and choose Quick Cover Page from the list. If you want to make one-time changes to the text, edit the text block that appears in the Cover Page section of the Send Fax dialog box.

FIGURE 16.5

The Select Cover Page dialog box

		Select Cover Page		
Available Cover Pages				
General	Thumbnail		Description	Search...
more covers				Restore
				Show All...

Selected Cover Page

Quick Cover Page

☐ Set as Default Cover Page

Edit...

OK   Cancel

Adding Attachment Files on the Fly

In earlier versions of WinFax, you had to convert files to fax attachment format before you could include them in a fax transmission. In WinFax Pro 4.0, you can select and attach files from the Send Fax dialog box, regardless of their format. If you frequently use the same files as attachments, it's still possible to use the Attachments command in the WinFax Fax menu to define attachment files, but it's no longer an absolute requirement.

To attach a file from the Send Fax dialog box, click on the Attach... button. When the Select Attachments dialog box in Figure 16.6 appears, either highlight a file from the list in the Attachment Library section and click on the Add To List button, or use the Attach File button to open a file selector and select a new file.

If the highlighted attachment file in the Send list is in WinFax format, a thumbnail view will appear next to the list. If the highlighted file is in some other format, you can see what the file contains by clicking the View... button. Use the Remove button to delete a file from the send list.

FIGURE 16.6

The Select Attachments dialog box

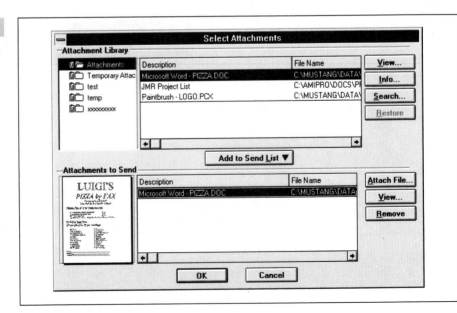

Using Multiple Dialing Prefixes on a Portable Computer

If you use WinFax to send faxes from your portable computer, you might have discovered that you need different dialing prefixes to connect your modem to different telephone lines. For example, when you place calls from your office, you might need to dial 9 to get an outside line before you dial the actual telephone number, but when you take the computer home for the weekend, you just dial the number. When you take the machine on a business trip, every hotel seems to have its own peculiar set of special dialing prefixes.

In earlier versions of WinFax, you had to type in the new prefix in the Send Fax dialog box every time you used your computer away from your original location. In WinFax Pro 4.0, you can define separate prefixes for your office and home, and for a third location (called "away").

When you change locations, select the User... command in the WinFax Setup menu to display the dialog box in Figure 16.7, change the information in the Present Location field, and type the new prefix in the Dial Prefix field or select it from the drop-down menu. The next time you send a fax, the Send Fax dialog box will contain the new prefix as part of the Number field. If the prefix is not correct, drop down the Dial Prefix menu in the Send Options dialog box to choose a different prefix.

Configuring WinFax for Credit Card Calls and Alternate Long-Distance Carriers

Most local telephone companies and major long-distance carriers in North America use the same standard format for customer-dialed credit card calls: dial zero, followed by the number you want to call, then wait for a tone and dial your credit card number (many telephone companies call them *calling card numbers*, but they work the same way). However, there are some long-distance services that use other arrangements. For

FIGURE 16.7

The User Setup
dialog box

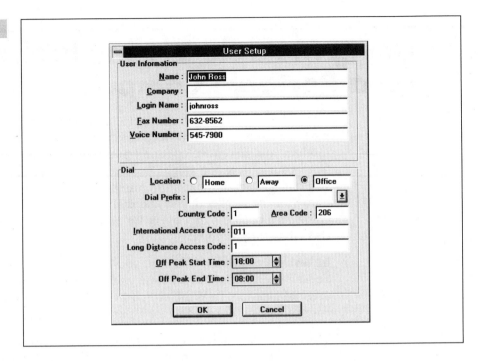

example, you might need to dial a local access number or a toll-free number to reach a discount carrier, and then dial a special access code before you dial the telephone number you want to reach. In WinFax Pro 4.0, you can automate this process, even if the dial sequence is not standard.

Every local telephone line in North America has been assigned to a primary long distance carrier, such as AT&T, Sprint, or MCI in the U.S, or Bell Canada or Unitel in Canada. When you dial 1 or 0, followed by the area code, you connect to your primary long-distance carrier. But it's also possible to place calls through any other long-distance carrier that serves your local telephone company by dialing 10*xxx*, where *xxx* is the carrier's access code. For example, you can reach AT&T by dialing 10288 (10-ATT). The access code for MCI is 10222. If you want to use an alternative long-distance carrier to send faxes, you can set up a credit card dialing sequence that automatically connects to the carrier of your choice.

To use the standard credit card dial sequence, select the Credit Card... command from the WinFax Setup menu. When the Dial Credit Card Setup dialog box in Figure 16.8 appears, type your credit card number, including the four-digit Personal Identification Number (PIN), which may not appear on your credit card.

The Dial Credit Card
Setup dialog box

```
┌─────────────────────────────────────────────────────────────┐
│  ┌──────────────────── Dial Credit Card Setup ──────────┐    │
│  Long Distance Service : [Custom                    ▼]  [ OK ]│
│  Credit Card Number :  [*********     ]            [ Cancel ] │
│                                                   [ Modify>> ]│
└─────────────────────────────────────────────────────────────┘
```

Keeping Your Credit Card Number Secure

WinFax masks your credit card number in dialog boxes, but if somebody else is using your computer, or if the computer is stolen, it's possible to charge calls to your credit card, even if the thief can't see the number. In addition, a sophisticated thief might find a way (which we won't describe here for obvious reasons) to extract your credit card number and use it for voice calls.

Therefore you should take these precautions to protect your credit card number:

- Don't type your credit card number in the Dial Credit Card Setup dialog box until you're about to place a credit card call.

- Delete your credit card number from the Dial Credit Card Setup dialog box after you're finished sending faxes on credit card calls.

- If you disregard this advice and keep your credit card number in the computer, notify the company that issued your credit card immediately if your computer is stolen.

Changing the Dial Sequence

The default Dial Sequence is the most common sequence for credit card calls. In most cases, you should leave it alone. But if your telephone credit card requires a non-standard dial sequence, click on the Modify>> button to view the expanded dialog box shown in Figure 16.9. To keep track of what you're doing, write the dial sequence on a worksheet, including the pauses to wait for a second dial tone or some other signal tone. When

FIGURE 16.9

The Expanded Dial
Credit Card Setup
dialog box

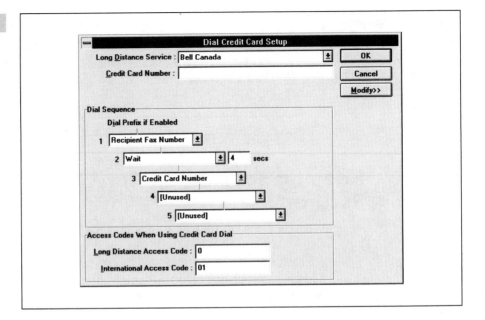

the expanded dialog box appears, split your dial sequence into pieces and place one piece in each numbered field in the Dial Sequence box. For example, if you have to dial a toll-free access number and then enter your account number and the number you want to call, select Service Access Number in Field No. 1 and type the toll-free number in the Service Access Number field. Choose Wait in Field No. 2, and set the duration to the number of seconds it takes to answer. In Field No. 3, choose Credit Card Number, and choose Wait in Field No. 4. Finally, choose Recipient Fax Number in Field No. 5.

If you want to use the Dial Credit Card Setup dialog box to place calls through an alternative long-distance carrier, set all five numbered fields to [Unused], and type the carrier's 10xxx access code in the Long-Distance Access Code field. To use an alternate long-distance carrier for overseas calls, type the 10xxx access code followed by 01 in the International Access Code field.

Transferring Files with WinFax

As Chapter 1 explains, fax communication converts images to electronic signals in a standard format, transmits them over a communication link, and converts the electronic signals back to images at the receiving end. That's about all you can do with most stand-alone fax machines, but when you use a fax modem or a late-model fax machine to send and receive computer files, it's possible to bypass the image conversion stages at either end and use the fax communication protocol to exchange computer data. WinFax 4.0 supports this kind of binary file transfer. Since binary file transfer doesn't need to bother with image conversion at each end, it's a lot faster than conventional faxing.

Why bother with binary file transfer? Why not use the same modem with a data communications program instead? If you're transferring files to another PC, error-correcting data transfer protocols such as ZMODEM are faster than faxing and at least as accurate as anything WinFax can do. But these are situations in which binary file transfer is your best method:

- Your modem has a top fax speed of 9600 bps, but a top data speed of only 2400 bps.

- You want to send data files directly from Windows applications, rather than starting a separate communications program.

- You want to send a fax to a machine that uses the Microsoft At Work architecture.

- You want to send a fax through a packet data network, such as MCI Mail.

- You want to send a fax as electronic mail over a local area network or through the Internet.

- You want to send the same document to several people, some of whom have machines that accept binary file transfer, and others of whom have conventional fax machines.

There are other reasons to use binary file transfer, but those are enough to show you that this technique has its advantages. On the other hand, you do need to know that your intended recipient is set up for binary transfer before you try to send something. If you try to send a file to an unsuspecting fax machine or modem, your transmission will probably fail. The initial release of WinFax Pro 4.0 supports generic binary file transfer (BFT), a proprietary WinFax4 protocol, and the Microsoft At Work architecture that will be available in many new office machines and in future versions of Windows. Later releases will probably support additional networks and protocols, but the basic technique for file transfer will remain the same.

Configuring Your Phonebook for File Transfers

Before you can send somebody a fax, you need to know that they have a fax machine or fax modem. Also, your recipient must be able to recognize one of the file transfer protocols before you try to send it. Therefore, until a single type of fax file transfer becomes the de facto standard, you'll either have to ask each person to whom you send faxes if they can receive file transfers, or assume they can't and fall back to conventional faxing.

As part of your phonebook configuration, you can record a list of file formats that each recipient can recognize. When you send a file to a recipient in one of those predefined formats, WinFax will automatically perform a binary transfer instead of converting it to a fax. If the recipient doesn't have a program to read that kind of file, WinFax will convert the document to a graphic image and send it as a conventional fax.

When you have learned that a person in your WinFax phonebook can use a file transfer protocol, use this procedure to change that person's record:

1. Open the WinFax Phonebooks window.

2. If it's not the current phonebook, click on the description of the phonebook that contains the record you want to change to display the contents of the phonebook in the item list.

3. Highlight and double-click on the entry in the item list for the listing you want to change.

4. When the Modify Record dialog box in Figure 16.10 appears, open the menu at the left side of the first line in the Connections box.

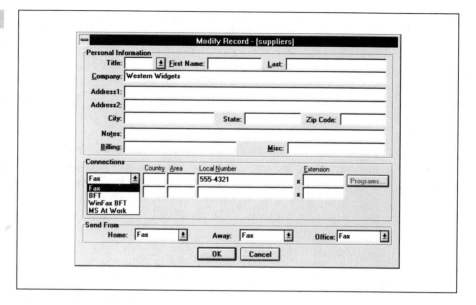

5. Select the protocol that this recipient is using.

6. When you choose BFT, WinFax4, or MS At Work, the Programs...
   button becomes usable. Click on this button to display the Programs Available At Recipient dialog box shown in Figure 16.11.

7. Choose the application programs that produce files this recipient can recognize from the Possible Programs list. This includes both the same programs that are on your computer and other programs that can read the same files. For example, most Windows word processors include filters that can convert from other word processors' file formats, so you should select amipro.exe and winword.exe, even if the recipient uses WordPerfect. This is equally true of databases and spreadsheets.

After you have highlighted all the programs you have in common in the Possible Programs list, click on the Add>> button. To delete a program from the Available list, highlight it and click on the <<Remove button. To delete everything in the Available list, click on <<Remove All.

You can send faxes to this person from Windows applications exactly the same way you send them to other recipients. However, WinFax will use the specified file transfer protocol to transfer the document, instead of converting it to a fax image. To send a data file or program file, click on the Attach... button in the Send Fax dialog box, and use the Attach File... button in the Attachments dialog box to open a file selector.

Configuring WinFax for Microsoft At Work

Microsoft At Work is a network interface for faxes, copiers, printers, and other office machines. You can send faxes and transfer files to Microsoft At Work-compliant devices through telephone lines or networks.

To send a fax or a file to a recipient's Microsoft At Work device, select MS At Work as the transmission type when you create a phonebook record for that recipient. In most cases, that's all you need to do, but if you want to use Microsoft At Work's security features, you can instruct WinFax to encrypt the data during transmission. If you do encrypt, the person receiving the fax will need to decode the data at the receiving end.

Microsoft At Work's security configuration is hidden under a couple of levels of dialog boxes. Click on the Options... button in the Send Fax dialog box, and then choose the At Work... button in the Send Options dialog box to display the dialog box in Figure 16.12.

FIGURE 16.12

The At Work Options
dialog box

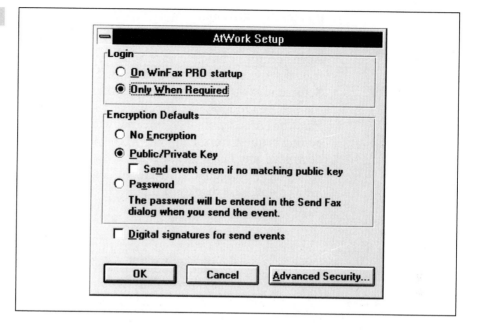

If you don't want to encode your transmission, make sure the No Encryption option is active. If you use the Public/Private Key Or Password option, be sure your recipient knows which method you used.

Using WinFax for Remote Polling

Even if you run WinFax on both an office computer and a portable, the people who send faxes to you probably send them all to your office fax number. When you're away from the office, you can use the remote polling feature of WinFax Pro 4.0 to call your office computer and download recently received faxes.

Setting Up the Remote System

Before you can dial into a computer to download faxes, the remote system must be set up for fax retrieval. To configure a system running Winfax for polling by a remote computer, select the Receive command in the Setup menu to display the Receive Setup dialog box shown in Figure 16.13.

At the bottom of the Receive Setup dialog box, make Enable Remote Retrieval active and click on the Setup... button. When the Remote Retrieval Setup dialog box appears, choose either None or Check Callers Password in the Secure Options box. If you select the Check Password option, type the Password used by the machine that will be polling you. If WinFax receives a polling request from a system whose Password does not match, it will not download any faxes. If you make the Mark Received Event As Read After Retrieval option active, the Receive Log will treat faxes that it has been polled for as if you have displayed them on your own machine.

FIGURE 16.13

The Receive Setup dialog box

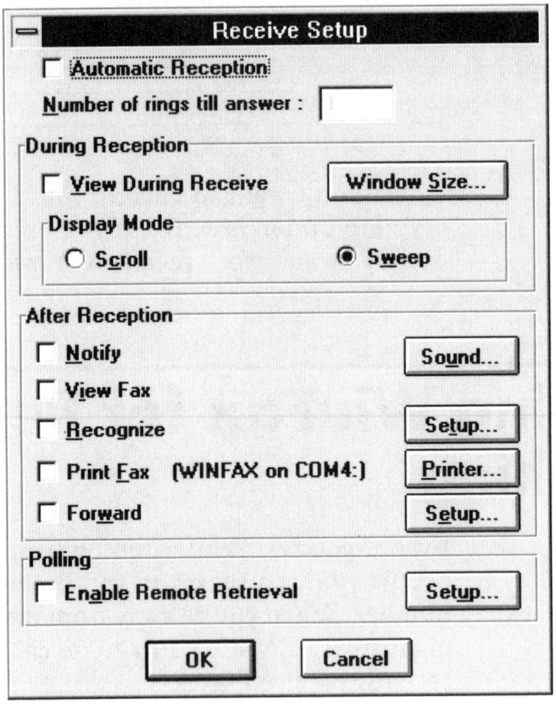

Downloading Faxes from a Remote System

To use the remote polling feature, select the Retrieve Remote... command from the Receive menu. When the Retrieve Remote dialog box appears, either click on the Select button to choose a name and telephone number from a phonebook, or type the name and fax telephone number of the computer from which you want to download faxes.

If the remote machine has been set to require password access, type the password in the Password field. If you need to include a dialing prefix before you dial the fax number, click on Use Prefix and either type the prefix in the field on the same line or select a prefix from the drop-down menu.

If you're expecting a fax from a particular source, and that's the only thing you want to retrieve from the remote system, type the Password of the originator (not the remote system from which you're retrieving the fax) in the Password field.

Automatically Forwarding Your Faxes

When you're away from your office for an extended period of time, it can be convenient to forward your faxes to another fax machine. You can configure WinFax to relay faxes from the same Receive Setup dialog box you used to enable remote retrieval.

To set up automatic forwarding, select Setup ➤ Receive.... From the Receive Setup dialog box, make Forward active at the bottom of the After Reception box, and click on the Setup... button in the same line to make the Forward Setup dialog box shown in Figure 16.14 appear.

The Forward Setup dialog box includes these options:

> **Forward To**   Type the telephone number of the machine to which you want to forward your faxes, or click the Select... button to choose a name and telephone number from a phonebook.

FIGURE 16.14

The Forward Setup
dialog box

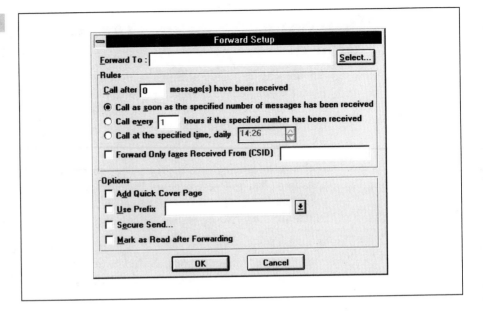

Rules   Choose an option to specify whether WinFax should forward faxes as soon as they arrive or on a time schedule. To limit forwarding to faxes from only one source, type the Password of the source in the Forward Only Faxed Received From field, and make the option active.

Options

> **Use Prefix**   Make this option active and choose the prefix from the drop-down menu if you need a dialing prefix to call the Forward To number.
>
> **Secure Send**   Make this option active to make sure you don't forward your faxes to the wrong destination. When the Secure Send dialog box shown in Figure 16.15 appears, type the Password of the fax machine or modem to whom you want to forward your faxes.
>
> **Mark As Read After Forward**   If this option is active, faxes that have been forwarded in the Receive Log will be

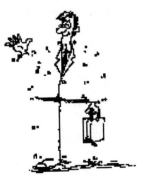

FIGURE 16.15

The Secure Send
dialog box

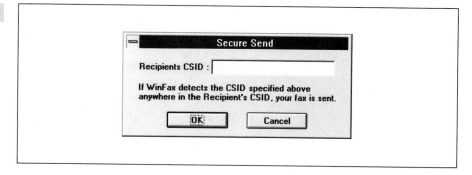

listed with the same status markings as those you have
read or printed locally.

Cover Page If you need to add your own cover sheet to for-
warded faxes, make this option active.

Adding Graphic Stamps to Your Annotations

It was possible to import a graphic image as an annotation in WinFax 3.0,
but you had to specify the path and file name of the graphic image file
every time you used it. In Version 4.0, you can create a list of frequently
used graphic annotations and import the images with a few mouse clicks.
This is an easy way to add your signature or initials to a fax, and to add
other images, such as a logo.

The Stamps... command in the Viewer's Annotate menu is visible after
you use the Show Annotation command. You can also enter the Stamps
command by clicking on the icon button with the cartoon rubber stamp
on it. The Signature dialog box shown in Figure 16.16 contains a list of
images. To add one of the images in the list to a fax, highlight the description and click on OK.

The new signature image appears in the Viewer as a graphic object. You
can use the annotation tools to move the signature, change the size, and

FIGURE 16.16

The Stamps dialog box

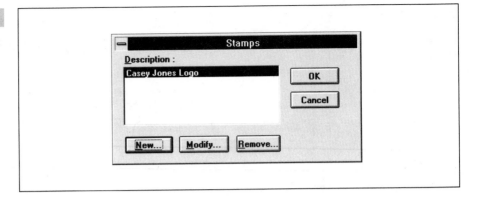

use the other annotation tools just as you use them on any other object. You can delete a stamp by clicking inside the object and selecting Edit ➤ Delete Object. Chapter 15 contains more detailed information about Annotation.

To add a graphic image to the Stamp list, use a graphics application program to create the image, and save it in one of these file formats:

- Windows Bitmap (.BMP)
- Tagged Image (.TIF)
- Paintbrush (.PCX)
- GEM Image (.IMG)
- GEM Metafile (.GEM)
- Mac Paint (.BIN)
- Encapsulated Postscript (.EPS)
- Graphic Interchange (.GIF)
- WinFax (.FXS)

After you have created the image file, start the Viewer and select the Show command in the Annotate menu. Next, select the Stamp command in the Annotate menu or click on the Stamp icon button. When the Stamp dialog box appears, click on the New... button to display the New Stamp dialog box shown in Figure 16.17.

Type a description of the image in the Name field, and either type the path and file name of the graphics file that contains the image in the File

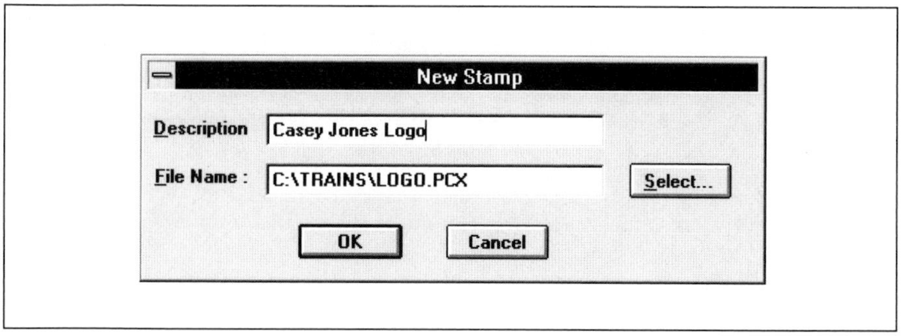

FIGURE 16.17

The New Stamp
dialog box

Name field, or click on the Select... button to open a file selector. When both fields are filled in, click on the OK button.

To change an existing stamp, highlight the description of that image in the Stamps dialog box and click on the Modify... button. The Modify Stamps dialog box is identical to the New Stamp dialog box, with the description and file name of the highlighted signature image already filled in. Edit the field you want to change, and click on OK.

To delete an image from the Stamp list, highlight the name of the image you want to remove and click on the Remove... button. When the program asks you to confirm the deletion, click on OK.

WinFax Pro 4.0 includes many changes and new features. Among other things, the newest version of WinFax Pro uses the familiar document windows layout that you know from other application programs, and there are features that make it easier to send faxes from portable computers, include attachment files with a fax, and automatically forward faxes after you receive them. In addition, you can use WinFax to send binary data files as well as images of documents. In this chapter, we've explained how to use all the new features in WinFax Pro 4.0. Using this information and the information in the preceding chapters, you should now be able to use WinFax to send and receive faxes and files and use all of the program's optional bells and whistles.

We hope it will be that simple. But a computer sometimes seems to have a mind of its own. In Part IV, you can find troubleshooting suggestions and explanations of error messages. With any luck, your system will work so well that you'll never need to open that part of the book.

Troubleshooting

WinFax isn't perfect. Like every computer program since Babbage, and like every communications technology since smoke signals, WinFax doesn't always do everything you expect it to do. This section contains descriptions of the most common problems you're likely to encounter while trying to install and use WinFax and suggestions for solving them. It also includes a list of WinFax error messages and an explanation of each message.

The information in this section applies to both WinFax Lite and WinFax Pro.

Symptoms
and Solutions

IT should be easy to send and receive faxes with WinFax, but sometimes something unexpected will happen. This chapter contains descriptions of some of the problems that may stand in the way of successful operation and suggestions for moving around them. It includes WinFax problems and related difficulties with fax modems and with Windows that can interfere with WinFax.

In the past, Delrina has issued new releases of WinFax without changing the version number and without telling existing users about them until they have a problem. As a result, there are at least three versions of WinFax Pro Version 3.0, and even more of WinFax Lite, including custom versions for individual modem types and crippled versions that only work with particular applications. You can download maintenance release patch files from Delrina's bulletin board in Toronto (416-441-2752) from the Delrina forum on CompuServe (GO DELRINA).

Many of the problems in this chapter apply to both WinFax Lite and WinFax Pro. If there's a difference between the two, the description of the problem identifies the version of WinFax in which it can occur and offers separate solutions for each.

Installing WinFax

These problems may appear while you're trying to install WinFax.

You get a General Protection Fault or Unrecoverable Application Error message during installation When WinFax sends test messages to identify your modem type, a few modems return data that WinFax can't handle. If this happens when you try to install WinFax, specify a different COM port (don't actually move the modem) and complete the installation. When installation is complete, follow these steps to change the COM port assigned to the WinFax printer driver:

1. Open the Printers dialog box in the Windows Control Panel.

2. Highlight the WinFax printer driver, and click on the Connect button (the Configure button in some versions).

3. When the Connect dialog box appears, select the COM port connected to your modem from the list of ports and click on OK.

Your fax modem isn't on the list in the setup dialog box, and you don't know if your modem is Class 1, Class 2, or Sendfax
To test your modem for fax operation, start your data communication program and enter this modem command:

ATI4

If the modem returns the word *Sendfax*, it's a Sendfax modem. If you're installing WinFax Lite, select the Generic Sendfax option in the fax device list. If you're installing Winfax Pro, select Sendfax from the Modem type list.

If the modem returns any other response to the ATI4 command, enter this command:

AT+FCLASS=?

The modem will return information about its class:

0,1	Modem is Class 1
0,2	Modem is Class 2
0,1,2	Modem is both Class 1 and Class 2

Install WinFax for a generic modem of the correct class. If your modem accepts both Class 1 and Class 2, you can choose either one.

Your fax modem doesn't respond properly WinFax loads a specific initialization string for each make and model of fax modem, with a set of initialization commands that are supposed to be optimized for that modem. In practice, the initialization strings don't always give you the best performance. If you have trouble sending or receiving faxes with the standard initialization string for your modem type, try reloading WinFax from the diskettes, but this time, select one of the generic modem types. If your modem supports both the Class 1 and Class 2 standards, try the modem-specific settings for both classes.

Sending Faxes

If you have difficulty sending faxes with WinFax, try one of these solutions.

The program appears to be running, but WinFax doesn't send faxes successfully There are several possible causes for this problem. Read the sections that follow to find a configuration that matches your own system:

You can't print from Windows to your own printer Your AUTO-EXEC.BAT file must include a SET TEMP= statement that specifies a valid directory for temporary files. If there is no SET TEMP= line in your AUTOEXEC.BAT file, or if it specifies a directory that does not exist, edit the file to include a valid SET TEMP= statement.

Make sure there is not a space at the end of the SET TEMP= line, or the Print command in all Windows applications will be grayed out.

You have removed and re-attached your modem since the last time you sent a fax Make sure the fax modem is attached to the correct COM port and that the modem is configured for that port address.

You're using a Stacker Co-processor board There's a memory conflict between WinFax Pro and the Stacker Co-Processor board. To eliminate this conflict, add this line to your SYSTEM.INI file, in the [386Enh] section:

EMM EXCLUDE=A000-EFFF

Your computer has a 386 processor WinFax uses up substantial amounts of your computer's resources. On a system with a 386SX processor, or one with a relatively slow CPU, you may not be able to keep up with the program's demands unless you make some changes to your Windows configuration.

To make additional system resources available to WinFax, add these lines to your SYSTEM.INI file:

Comboosttime=30
Com*x*autoassign=0

Com*x*Buffer=1024
Com*x*Protocol=XOFF

In place of Com*x*, use the port number connected to your fax modem. For example, if your fax modem is on COM2, the second line will be Com2autoassign=0.

After you change the SYSTEM.INI file, restart Windows and try sending a fax. If you still have trouble sending, open the 386 Enhanced section of the Windows Control Panel and change the minimum time slice to 30 milliseconds.

If the problem still occurs, try running Windows in Standard Mode.

You're trying to send a fax from Word for Windows   If WinFax consistently fails to send faxes from Word for Windows, select the Print command in Word's File menu, and select the Options button. When the Options dialog box appears, make sure "Draft Output" in the Printing Options box is *not* active.

You're using an Intel SatisFAXtion 100, 300, or 400e modem   The installation routine for these modems loads the Intel CAS-emulation software, but WinFax treats them as Class 1 modems.

To use WinFax with one of these modems, disable the CAS-emulation drivers by placing **REM** at the beginning of these lines in your AUTOEXEC.BAT file:

C:\FAX\CASMODEM.EXE
C:\FAX\FAXPOP.EXE

In addition, you must change your WinFax installation. Run SETUP from the WinFax program diskettes and select the Change Modem option. When the list of modem types appears, choose the Generic Class 1 modem type.

Your fax is garbled or distorted at the receiving end   This problem may be caused by a noisy telephone line, but if *all* of your faxes are garbled, you may have a conflict with your Windows screen saver, which is looking for IRQ activity on your communication ports. To avoid this problem, you must either disable the screen saver's IRQ monitoring, or turn off the screen saver when you send and receive faxes. For example, in After Dark, you can turn off system IRQ activity in the Setup dialog box.

It's not possible to disable IRQ monitoring in the Windows 3.1 screen saver, so you must set the Windows screen saver to None. The Screen Saver control is in the Desktop section of the Windows Control Panel.

If you also get garbled output when you print from other Windows applications on your own printer, you may be having a problem with a font program such as Adobe Type Manager.

The computer reboots after you send a fax You probably have a hardware IRQ conflict. To identify the conflict, either close Windows or open a DOS window, and type **MSD** to start the Microsoft Diagnostic program. Click on IRQ Status to display a complete list of interrupts.

Look in the Detected column to find the COM port connected to your modem. If another device is also using that IRQ number, you have a potential conflict. To eliminate the conflict, reinstall either your modem or the other device using the IRQ, and assign the device to a different IRQ number.

WinFax Lite sends faxes at 4800 bps, even though you're using a 9600 bps or 14,400 bps modem You have WinFax configured for the wrong data speed. Select the Setup command in the WinFax Administrator, and click on the More button to open the Fax/Modem Setup dialog box. The Max Tx Rate field should be set to 9600/14.4K.

WinFax gives you "No carrier" messages when you try to send faxes overseas It can take more time to complete an international call than a call within your own country. To increase to 60 seconds the amount of time your modem waits for a carrier, add this string to the Init String in the Fax/Modem Setup dialog box:

    ATS7=60

Your faxes have two header lines If you use WinFax with an Intel CAS fax modem, the modem's firmware adds its own header line. If you discover you're sending faxes with two headers, use the WinFax Program Setup dialog box to delete all information from the Header section.

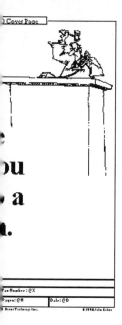

Encapsulated PostScript (EPS) graphic images in your document are missing from the fax WinFax before Version 4.0 does not recognize graphic images in PostScript format. As a result, WinFax will leave holes in a document that contains EPS graphics in place of the images. This can be a serious problem if you're trying to send an existing document as a fax.

The best way to work around this problem is to avoid using EPS graphics in documents you plan to send as faxes. When you use a graphics program such as CorelDRAW to create images, save them in TIF formats. If an existing document has EPS graphics, the best you can do is to convert the images to another format.

All transmitted pages from Freelance Graphics are blank When you print Freelance Graphics to WinFax, you must use the Print Setup dialog box (in the File menu) to configure the pages as either speaker notes or handouts.

You can't select WinFax as your printer in Microsoft Works, Microsoft Publisher, or Microsoft Money Works, Publisher, and Money are all part of Microsoft's Solution Series, and all three use the same printer selection technique. They don't recognize a device connected to a COM port as a printer unless it is the default printer. To send a fax from one of these programs, you must use the Printer section of the Windows Control Panel to make WinFax the default printer.

Problems sending from OS/2 WinFax can only use 1000 interrupts per second when it operates under OS/2, but that's not enough to send faxes at 9600 baud. To fix this problem, contact IBM for updated OS/2 communication drivers, and use these DOS settings:

```
IDLE_SENSITIVITY=100
IDLE_SECONDS=60
COM_HOLD=ON
```

This problem is less likely to occur if you keep WinFax in the foreground when you send a fax.

WinFax uses different fonts from the ones you specified Some laser printers and other devices use fonts that are resident in the print device's memory, instead of downloading them from the computer with each document. If you try to use one of those fonts with WinFax, Windows substitutes a different font, which may be *quite* different from the font you are expecting.

To fix this problem, make WinFax the current printer in the application from which you want to send a fax, and select fonts from the current set of available fonts.

WinFax asks for a fax telephone number in the middle of a multipage document When you create a Word for Windows document with pages in both portrait and landscape orientation, it starts a new document every time it shifts from one orientation to the other. As a result, the Send Fax dialog box appears each time the orientation changes. This is a nuisance, but it's not a bug. Complete the dialog box each time it appears (remember not to send additional cover sheets).

Ventura Publisher does not list WinFax in the Printer Setup list If you have a large number of Windows printer drivers installed, Ventura only recognizes the first six, even if some are inactive. Follow these steps to move WinFax to a higher position on the list:

1. Open the Control Panel from the Main group in the Windows Program Manager

2. Click on the Printers icon to display the Printers dialog box.

3. Remove one or more printer drivers that appear on the list ahead of WinFax until WinFax is among the first six.

4. If you wish, reinstall the printer drivers you removed.

Graphics are missing when you send faxes from Word for Windows If you try to send a Word for Windows document with graphic images in segments larger than 64 KB, some older versions of WinFax won't accept them. If this happens to you, either use the program that originally created the graphic image to reduce the resolution, or convert the graphic image from color to black and white, using either the Windows Paintbrush program or a graphics application.

You can't open the PackRat phonebook The WinFax Pro maintenance patch release does not properly modify the [WfxPbLinks] section of your WIN.INI file. To fix the problem, if you're using PackRat version 3.x or 4.1, change the [WfxPbLinks] section from:

```
[WfxPbLinks]
PackRat=1,packrat.exe,PackRat,System,PACK3
PackRat4.0=1,packrat.exe,PackRat,System,PACK3
```

to

```
[WfxPbLinks]
PackRat=1,c:\packrat.exe,PackRat,System,PACK3
PackRat4.0=1,packrat.exe,PackRat,System,PACK3
```

Or if you're using PackRat 4.0, change it to the following:

```
[WfxPbLinks]
PackRat=1,packrat.exe,PackRat,System,PACK3
PackRat4.0=1,c:\packrat.exe,PackRat,System,PACK3
```

The PackRat phonebook appears to be blank WinFax sorts PackRat phonebook records alphabetically by name and places blanks at the beginning of the list. If you have five or more PackRat client records with a blank Name Field, the Send Fax dialog box will display the visible part of the phonebook as blank. To avoid this problem, add an entry (such as the Company Name) to the Name field of each PackRat client record.

Receiving Faxes

This section contains solutions to problems that may occur when you are receiving faxes through WinFax.

WinFax does not answer incoming calls, even though Automatic Reception is active You may have a conflict with a Windows screen saver, which is looking for IRQ activity on your communication ports. You must either disable the screen saver's IRQ monitoring, or turn off the screen saver when you send and receive faxes. For example, in After Dark, you can turn off system IRQ activity in the Setup dialog box.

It's not possible to disable IRQ monitoring in the Windows 3.1 screen saver, so you must set the Windows screen Saver to None. The screen saver control is in the Desktop section of the Windows Control Panel.

WinFax prints incoming faxes on half of the page WinFax prints fax images as 204×196 dots per inch graphics. If the printer is not set to a square aspect ratio, it distorts the image and prints on only half of each page. To fix this problem, change your printer's aspect ratio for Windows to a square resolution, such as 180×180 dpi.

Problems receiving from OS/2 WinFax uses a maximum of 1000 interrupts per second under OS/2, but it needs more than that to receive faxes at 9600 baud. To fix this problem, contact IBM for updated OS/2 communication drivers and use these DOS settings:

```
IDLE_SENSITIVITY=100
IDLE_SECONDS=60
COM_HOLD=ON
```

You may be able to eliminate this problem by keeping WinFax in the foreground during reception.

The optical character recognition (OCR) feature won't work because of insufficient memory, even though you believe you have a total of at least 4.5 megabytes in RAM and the Windows swap file The WinFax OCR feature requires a minimum of 4.5 megabytes of combined available memory in RAM and the Windows swap file. To display the amount of free memory currently available on your system, select the About command from the Windows Program Manager Help menu.

To create more free memory, press Ctrl-Esc to open the Windows Task List, and close any other Windows applications that you don't need at the moment. If you are running any memory resident DOS programs (including the WinFax DOS TSR program), edit your AUTOEXEC.BAT and CONFIG.SYS files to remove them, and restart your computer.

If you are not using a permanent Windows swap file, you should create one from the Windows Control Panel.

If OCR still doesn't run, edit the [AnyFax] section of your WIN.INI file to change the line that looks like this:

```
Memsize_300=1800
```

so that it looks like this:

Memsize_300=1000

and add this line to the [AnyFax] section:

TempDiskSpace=800

Printing Faxes

This section contains solutions to problems that can occur when you print faxes from WinFax.

There are white gaps in printed copies of faxes The Image Viewer in some early versions of WinFax skips part of the image when it prints from 50% and 25% views, which makes the printed copy look as if it has been chopped up. To print a clean copy, change the view to 100%.

Delrina has fixed this problem in more recent versions of WinFax Pro, so you can avoid the problem by updating your software to Version 3.0 or Version 4.0.

Problems while Using the WinFax Administrator

This section contains solutions to problems you may encounter in the WinFax Administrator.

The screen font in WinFax icon buttons and in the Send and Receive logs overflows the icons and text boxes Normally, WinFax uses the Sans Serif screen font for icon buttons and text boxes. If some other program changes the [Fonts] and [FontSubstitutes] section of your WIN.INI file, WinFax may use some other screen font.

Follow these steps to return to the original screen fonts:

1. Use a text editor to display your WIN.INI file, which is situated in the WINDOWS directory.

2. Look for any line that begins like this:

   Helv=

3. Place a semicolon (;) at the beginning of each Helv= line to turn it into a remark line.

4. Look in the [Fonts] section for any line that looks like this:

   MS Sans Serif 8,10,12,14,18,24, (VGA res)=SSERIFE.FON

   Make sure this line does not start with a semicolon. If the line is missing, add it.

5. Make sure this line is in the [FontSubstitute] section of WIN.INI:

   Helv=MS Sans Serif

6. If you made any changes to WIN.INI, save the file.

7. Restart Windows.

8. Use the File Manager to look in the WINDOWS\SYSTEM\ subdirectory for the SSERIFE.FON file. If it's missing, use the Fonts section of the Windows Control Panel to install it.

Conflicts between WinFax and Other Applications

Look in this section for information about using WinFax and other applications at the same time.

You can't start a data communications program when WinFax Automatic Receive is active When Automatic Receive or the WinFax DOS TSR program is active, WinFax controls the COM port connected to your fax modem and it won't allow access to the port by most other applications. If you have trouble starting a data communications program, turn off the WinFax Auto Receive option or remove the TSR program, and then enter the command to start the data communications program again. When you're done with your data communication session, remember to turn the WinFax auto receive function back on.

WinFax conflicts with Calendar Creator WinFax and Calendar Creator 1.0 both use some of the same Windows resources, so the two programs may interfere with each other. This problem has been solved in Calendar Creator 2.0; if you have trouble using Calendar Creator, upgrade to the latest version.

In this chapter, we have covered some of the problems you may encounter with WinFax. In the next chapter, we'll discuss some of the error messages that appear in WinFax.

Error Messages

THIS chapter contains a list of error messages that may appear in WinFax Lite and WinFax Pro and an explanation of each message.

WinFax Error Messages

Since WinFax works closely with other Windows applications and with Windows itself, you may see error messages that were generated by other programs. Look in the manual for Windows and the other applications (or in your library of SYBEX books about them!) to find information about those messages.

Application Error—floating point error: invalid This message appears when you try to include a cover page when you send a fax from Drafix Windows CAD 2.1. To work around the problem, save the Drafix file as a WinFax attachment and send it from the WinFax Administrator. If you don't need a cover page, you can send faxes directly from Drafix Windows CAD 2.1.

Cannot Attach File (WinFax Lite) This message appears when you try to attach a fax file that you have previously rotated to a different orientation from the rest of the same fax. If you see this error message, use the Image Viewer to rotate the attachment file and try to send it again.

Can't find XXXX.DLL (WinFax Pro) This message appears when you try to start WinFax and the program is unable to find one of the files it needs. If you see this message, look for the named file in your WinFax directory. If it's there, make sure the WinFax directory is in your DOS search path by entering this DOS command:

```
PATH
```

You should get a list of the directories in your DOS search path. If it does not include the WinFax directory, edit your AUTOEXEC.BAT file to add it. Remember that you must include the drive letter for each directory, so the AUTOEXEC.BAT command should look something like this:

```
PATH C:\;C:\DOS;C:\WINDOWS;C:\WINFAX
```

Error enabling source (WinFax Pro) This message appears when you try to use one of the WinFax Scan commands and your scanner is not compatible with the TWAIN standard or the scanner software is not properly configured.

If you see this message, find the manual for the scanner software and confirm that it is TWAIN-compatible. It probably is, because WinFax does not accept Scan commands unless it detects a TWAIN scanner. Make sure the scanner is configured for black and white one bit per pixel images, with 200×200 dpi resolution. The image type must be set to Line Drawing, Black And White Drawing, or Line Art. WinFax won't work with color or gray-scale settings.

Error initializing attachments database-1 This message appears when you try to use WinFax to save a WinFax phonebook as a fax attachment file. If you see this message, either export the phonebook to an ASCII text file and use a word processor or text editor to send the file as a fax, or disable the Add To Attachment list option in the Save To File dialog box.

Error locating group member record (WinFax Pro) This message appears when you try to send a fax to a recipient group and one or more records in that group has a first name with exactly 15 characters.

If you see this message, edit the recipient records in your phonebook to reduce the maximum size of the first name field to 14 characters.

Error locating hardware (WinFax Lite) This message appears when you start the WinFax Lite Administrator program and your fax modem is turned off or is not connected to the correct COM port. It can also appear if you try to start WinFax when another Windows or DOS application is using the modem port.

If you see this message, make sure the modem is properly connected and the modem's power switch is on (if you're using an internal modem, it turns on whenever you turn on the computer). If none of the lights on the front of the modem are on, there may be no power to the modem. Press Ctrl-Esc to display a list of currently active Windows applications. If any of the active programs use the modem port, either close those applications or wait for them to complete their work, and then try to start WinFax again.

Error opening cover page file (WinFax Pro) This message appears when you try to include a cover page when you send a phonebook as a fax. If you see this message, either try sending the phonebook without a cover page or export the phonebook to an ASCII text file, and use a word processor or text editor to send the file as a fax.

Error opening source (WinFax Pro) This message appears when you try to use one of the WinFax Scan commands and your scanner is not compatible with the TWAIN standard or the scanner software is not properly configured.

If you see this message, find the manual for the scanner software and confirm that it is TWAIN-compatible. It probably is, because WinFax does not accept Scan commands unless it detects a TWAIN scanner. Make sure the scanner is configured for black and white one bit per pixel images, with 200 × 200 dpi resolution. The image type must be set to Line Drawing, Black and White Drawing, or Line Art. WinFax won't work with color or gray-scale settings.

Error running WinFax PRO (WinFax Pro) This message appears when you try to send a fax directly from another Windows application, and the WinFax printer driver is unable to start the FAXMNG.EXE program.

If you see this message, confirm that the FAXMNG.EXE file is in your WinFax directory. If it's missing, reinstall WinFax from the diskettes. If it's there, make sure the WinFax directory is in your DOS search path by entering this DOS command:

    PATH

You should get a list of the directories in your DOS search path. If it does not include the WinFax directory, edit your AUTOEXEC.BAT file to add it. Remember that you must include the drive letter for each directory, so the AUTOEXEC.BAT command should look something like this:

```
PATH C:\;C:\DOS;C:\WINDOWS;C:\WINFAX
```

Error spawning WinFax Administrator (WinFax Lite) This message appears when you try to send a fax with WinFax Lite, but the WinFax Administrator application file, FAXMNG.EXE, is not in the directory you specified when you installed WinFax Lite.

If you see this message, reinstall WinFax Lite from the diskettes.

FAXMNG.EXE caused a General Protection Fault in module WFXPB.DLL (WinFax Pro) This message appears when you try to scroll through a PackRat phonebook from the Send Fax dialog box. It may also appear when you open the Send Fax dialog box.

If you see this message, change the PackRat phonebook format to make the fax number one of the first four number fields in the client record. In order to eliminate this problem, you can install the WinFax WFX304 maintenance release patch, which is available free from the Delrina Bulletin Board (416-441-2752) and the Delrina Forum on CompuServe.

General Protection Fault (WinFax Lite and WinFax Pro) This message appears when WinFax (or any other Windows application) tries to write to a memory space where it doesn't have access, which corrupts the code already written to that piece of memory. You may see a General Protection Fault (GPF) if Windows fails to erase a temporary file.

If you see this message, click on the Ignore button. If the message appears again, keep clicking on Ignore or Close. When the message finally goes away, exit WinFax and then exit Windows. If WinFax won't close or the computer locks up, use the Windows 3.1 local reboot command by pressing Ctrl-Alt-Del, and follow the instructions on your screen.

When you're at the DOS prompt, type **SET** to find out where Windows places temporary files. Change to the directory identified in the TEMP= line, and type this command:

```
DEL *.TMP
```

After you delete the temporary files, restart Windows and repeat the sequence that caused the GPF. If it happens again, follow the same set of steps to bail out of your application and Windows, and run CHKDSK /F from the DOS prompt.

If a General Protection Fault occurs at all, it will probably occur immediately after you install WinFax or some other program, or after you make some kind of change to your system. Dr. Watson is a diagnostic program included with Windows that creates a DRWATSON.LOG file when an error happens. If you still get a General Protection Fault after deleting .TMP files and running CHKDSK /F, run Dr. Watson, and start WinFax again. After the error occurs, you should be able to find the problem in the DRWATSON.LOG file.

Image contains invalid data This message appears when you try to print or view a page of a received fax message, but the image file is either a failed event or a corrupted event, or WinFax doesn't recognize the file format. You will also see this message if you try to read a file with a .PCX, .FXS, .FXD, or .FXR file extension, but it's not actually an image file.

WinFax could be conflicting with your Windows screen saver, which is looking for IRQ activity on your communication ports. To successfully view a fax, either disable the screen saver's IRQ monitoring, or turn off the screen saver when you send and receive faxes. For example, in Afterdark, turn off system IRQ activity in the Setup dialog box.

It's not possible to disable IRQ monitoring in the Windows 3.1 screen saver, so you must set the Windows screen saver to None. The screen saver control is in the Desktop section of the Windows Control Panel.

Invalid Image File (WinFax Lite) This message appears when you try to display or print a page of a received fax that contains corrupted data. The corruption is probably caused by a transmission error. If you see this message, it's not possible to recover this page. The only way to obtain the corrupted pages is to ask the sender to fax another copy to you.

No communication from fax/modem You may have a conflict with a screen saver. Either disable the screen saver's IRQ monitoring, or if that's not possible, turn off the screen saver.

Not enough disk space to install (WinFax Pro) This message appears during installation when your C: drive has fewer than 600K of free disk space. The WinFax setup program requires this much space for the temporary WINFAX.T directory that it uses during installation.

If you see this message, run CHKDSK /F or a disk analysis and repair utility such as Norton Disk Doctor in The Norton Utilities or DiskFix in

PC-Tools to find and eliminate lost clusters. If you still don't have 600K free, delete enough unnecessary files to free up the space. Among the files you can safely eliminate are .TMP files, any files in the TEMP directory, the WIN386.SWP file, and any wallpaper (.BMP) and sound (.WAV) files that you don't ever use.

If you have more than one hard drive, try moving files or application files to a different drive in order to create more free space on the C: drive.

If you don't have 600K free on your C: drive, it's possible to modify the WinFax setup program to load the program to a different drive, but it's probably not a good idea, since you will almost certainly need at least that much space for image files.

Reception error _NN_ This message appears when WinFax tries to receive a fax, but there was some kind of interference in the incoming signal before WinFax got it. In general, reception errors occur when the fax call comes through a noisy telephone line, or when you have a problem with your modem or with the fax machine that originated the call.

For specific details about the cause of a reception error, find the error code in the list at the end of this chapter.

If you get repeated reception errors, the problem may be caused by telephone circuit problems. If you were able to receive an identification from the sender, or if you received one or more pages before the line went down, you might want to place a voice call to let the sender know about the problem, and suggest that she either wait half an hour and try again or that she use a different long-distance carrier by adding a 10xxx prefix when dialing your telephone number. For example, if she normally uses MCI as her long-distance carrier, suggest she try dialling 10288, followed by your area code and number, to place the call through AT&T lines. If she normally uses AT&T, try 10222 for MCI or 10333 for Sprint.

Source and destination directories are the same. No files will be copied or deleted. This message appears when you try to save log events to an archive file where the event files already exist. You must store archive files in a different directory from the one that holds the original log events.

The COM port is in use by another application. Make this port available to WinFax? This message appears when you try to start the WinFax automatic receive function, but another Windows communications application is already using your modem port. For example, if you are running a data transfer to another computer in one window, you can't take over the modem port to wait for incoming faxes. But of course, since the modem is in use, the telephone line is busy and no incoming faxes can reach your modem.

If you see this message, click on No to return to the other application or Yes to interrupt the other application to take over the modem port. The other application continues to run, but it no longer exchanges data with the modem.

Timeout waiting for POKE acknowledgment (WinFax Pro) This message appears in some versions of WinFax Pro 3.0 when you are trying to read a WinFax phonebook from a Personal Information Manager, because WinFax only allows fifteen seconds to load the PIM phonebook.

If you see this message, you can change the timeout setting to 120 seconds by installing the release patch WFX3PR.ZIP. You can download this patch from Delrina's bulletin board at 416-441-2752, or from the Delrina forum on CompuServe. If you're using a version of WinFax Pro 3.0 dated earlier than April 21, 1993, download and install the WFX304.ZIP patch file at the same time.

Transmission Error NN This message appears when WinFax tries to send a fax, but is unable to complete the transmission. In general, transmission errors occur when the fax call goes through a noisy telephone line or when you have a problem with your modem or the fax machine that tries to receive the call.

For specific details about the cause of a transmission error, find the error code in the list at the end of this chapter.

If you get transmission errors, the problem may be because of telephone circuit problems. Try waiting half an hour or more and then place the call again. If the fax you want to send is urgent, try using a different long-distance carrier by adding a 10*xxx* prefix to your telephone number. For example, if you normally use MCI as your long-distance carrier, try dialing 10288, followed by the area code and number, to place the call through

AT&T lines. If you normally use AT&T, try 10222 for MCI or 10333 for Sprint.

Twain error NN This message appears when you try to use a WinFax Scan command but your scanner is not compatible with the TWAIN standard or the scanner software is not properly configured.

If you see this message, find the manual for the scanner software, and confirm that it is TWAIN-compatible. It probably is, because WinFax does not accept Scan commands unless it detects a TWAIN scanner. Make sure the scanner is configured for black and white one bit per pixel images, with 200 × 200 dpi resolution. The image type must be set to Line Drawing, Black And White Drawing, or Line Art. WinFax won't work with color or gray-scale settings.

Unrecoverable application error This message appears when an error in WinFax or another application causes the application to stop running. If you see this message, click on OK to close the application and then restart Windows. If that doesn't work, reboot the computer.

WinFax cannot communicate with the fax/modem. Ensure that the fax/modem is switched on and connected to the serial port. This message appears when WinFax can't find your modem. If you see this message, make sure the modem is turned on and that the cable from your serial port to the modem is properly connected at both ends. If you're using an internal modem, shut down any applications currently running, exit Windows, and then use the power switch to turn your computer off and then on again (don't use Ctrl-Alt-Del). If that doesn't work, make sure both WinFax and your modem are configured correctly.

WinFax program failure. Please see manual for solutions. This message appears when you don't have enough free memory or disk space to allow WinFax to run, or when WinFax is unable to obtain access to a file that it requires in order to operate properly.

If you see this message, try removing unnecessary files from your hard disk, and disable any memory-resident programs that are not really necessary. Use the About Program Manager... command in the Windows Program Manager Help menu to find out how much memory is currently available.

WinFax supports 1 bit/pixel (black and white) images only This message appears when you try to scan a color or gray-scale picture into WinFax. If you see this message, make sure your scanner software is configured for 200 dpi black and white line drawings.

Class 1/2 Transmission and Reception Error Codes

When WinFax reports a Transmission Error or a Reception Error, it includes a code number that identifies the specific source of the error.

1 (Class 1)	Ring detect without successful connection.
2 (Class 1/2)	User aborted transmission/reception operation.
3 (Class 2)	No loop current.
4 (Class 1/2)	Phone line busy.
5 (Class 2)	No dial tone. Modem error. Possibly not installed correctly. No modem response. Possibly not turned on.
8 (Class 1)	Modem error while setting up call.
10 (Class 1)	Polling request received (not supported).
11 (Class 1/2)	Remote station did not answer the call.
20 (Class 1/2)	Modem error while sending DCS (digital command signal) or TCF (training check). Refer to Phase B of the T.30 protocol.
21 (Class 1/2)	Unable to support the send/receive parameters.
22 (Class 2)	COMREC (command received) error.

23 (Class 1)	Remote station issued disconnect signal.
24 (Class 2)	COMREC (command received) invalid.
25 (Class 1/2)	RSPEC (response received) error.
26 (Class 2)	DCS (digital command signal) received, not recognized.
27 (Class 1/2)	FTT (failure to train) received at all data rated.
28 (Class 1)	Protocol error after sending TCF (training check).
28 (Class 2)	RSPEC (response received) invalid.
40 (Class 1/2)	User aborted transmission operation.
43 (Class 2)	Data underflow. Processor may be overloaded.
50 (Class 1/2)	Modem error after data transfer. Modem error while sending MPS (multipage signal), EOP (end of procedure), or DCN (disconnect).
51 (Class 2)	RSPEC (response received) error.
52 (Class 1/2)	Frame error encountered while sending MPS (multipage signal).
53 (Class 1/2)	Protocol error when sending MPS (multipage signal). Unknown response.
54 (Class 1/2)	Frame error in response to sending EOP (end of procedure).
55 (Class 1/2)	Protocol error encountered when sending EOP (end of procedure).
56 (Class 2)	No response to EOM (end of message).
57 (Class 2)	Invalid response to EOM (end of message).
58 (Class 2)	Protocol error.

70 (Class 1/2)	Modem error setting modem data rate.
71 (Class 2)	RSPEC (response received) error.
72 (Class 2)	COMREC (command received) error.
73 (Class 2)	Timed out. Page not received.
74 (Class 2)	Timed out after EOM (end of message) received.
90 (Class 1)	Unable to write data to disk.
100 (Class 1/2)	Protocol or modem error while receiving MPS (multipage signal), EOM (end of message), EOP (end of procedure), PRI-Q (procedure interrupt), or DCN (disconnect).
101 (Class 2)	RSPEC (response received) invalid.
102 (Class 2)	RSPEC (response received) invalid.
103 (Class 2)	Protocol error.

CAS Modem Codes

0002 (Subcode: 2)	Bad scanline count.
0003 (Subcode: 3)	Page sent with errors.
0004 (Subcode: 4)	Receive data lost.
0005 (Subcode: 5)	Invalid or missing logo file.
0006 (Subcode: 6)	File name does not match NSF (non-standard format) header.
0007 (Subcode: 7)	File size does not match NSF (non-standard format) header.

0101 (Subcode: 1)	Invalid function number.
0105 (Subcode: 5)	Access denied.
0106 (Subcode: 6)	Invalid handle.
0200 (Subcode: 0)	Multiplex handler failed. Function failed. Data not sent.
020A (Subcode: A)	Invalid command parameter. Function failed. Data not sent.
020B (Subcode: B)	Cannot uninstall Resident Manager. Function failed. Data not sent.
020C (Subcode: C)	File already exists. Function failed. Data not sent.
0201 (Subcode: 1)	Unknown command. Function failed. Data not sent.
0202 (Subcode: 2)	Event not found. Function failed. Data not sent.
0203 (Subcode: 3)	Attempted to find next before find first. Function failed. Data not sent.
0204 (Subcode: 4)	No more events. Function failed. Data not sent.
0207 (Subcode: 7)	Invalid queue type. Function failed. Data not sent.
0208 (Subcode: 8)	Bad control file. Function failed. Data not sent.
0209 (Subcode: 9)	Communication board is busy. Function failed. Data not sent.
0280 (Subcode: 80)	Unknown task type. Function failed. Data not sent.
028A (Subcode: 8A)	Hardware command set error. Function failed. Data not sent.
028B (Subcode: 8B)	Bad NSF (non-standard form) header file. Function failed. Data not sent.

0281 (Subcode: 81)	Bad phone number. Function failed. Data not sent.
0282 (Subcode: 82)	Bad PCX file header. Function failed. Data not sent.
0283 (Subcode: 83)	Unexpected EOF (end of file). Function failed. Data not sent.
0284 Subcode: 84)	Unexpected disconnect. Function failed. Data not sent.
0285 (Subcode: 85)	Exceeded maximum dialing retries. Function failed. Data not sent.
0286 (Subcode: 86)	No files specified for send event. Function failed. Data not sent.
0287 (Subcode: 87)	Communication board timeout. Function failed. Data not sent.
0288 (Subcode: 88)	Received more than 1023 (maximum) pages of data. Function failed. Data not sent.
0289 (Subcode: 89)	Manual connect posted too long ago. Function failed. Data not sent.
0302 (Subcode: 2)	File not found. DOS error. Data not sent.
0303 (Subcode: 3)	Path not found. DOS error. Data not sent.
040A (Subcode: A)	No dial tone.
040B (Subcode: B)	Fax error. Invalid response from remote unit after sending data.
040D (Subcode: D)	Fax error. Phone line dead or remote unit disconnected.
040E (Subcode: E)	Fax error. Timeout while waiting for secondary dial tone.
0401 (Subcode: 1)	Fax error. Remote unit not Group 3 compatible.

0402 (Subcode: 2)	Fax error. Remote unit did not send its capabilities.
0403 (Subcode: 3)	Fax error. Remote unit requested disconnect.
0404 (Subcode: 4)	Fax error. Remote unit is not capable of file transfers.
0405 (Subcode: 5)	Fax error. Exceeded retrain or fax resend limit.
0406 (Subcode: 6)	Fax error. Line noise. Local and remote unit do not agree on a bit rate.
0407 (Subcode: 7)	Fax error. Remote unit disconnected after receiving data.
0408 (Subcode: 8)	Fax error. No response from remote unit after sending data.
0409 (Subcode: 9)	Fax error. Capabilities of remote unit are not compatible.
041F (Subcode: 1F)	Fax error. Unexpected EOF (end of file) while receiving.
0411 (Subcode: 11)	Fax error. Invalid command from remote after receiving data.
0415 (Subcode: 15)	Fax error. Tried to receive from incompatible hardware.
045C (Subcode: 5C)	Fax error. Received data overflowed input buffer.
045D (Subcode: 5D)	Fax error. Remote hardware unexpectedly stopped sending data.
045E (Subcode: 5E)	Fax error. Remote hardware did not send any data.
045F (Subcode: 5F)	Fax error. Remote hardware did not send any data.
0463 (Subcode: 63)	Fax error. Cannot get through to remote unit.
0464 (Subcode: 64)	Fax error. User cancelled event.

APPENDIX

A

Windows Dynamic Data Exchange Specifications for WinFax

THE DDE protocol is a set of rules that Windows applications use to exchange commands and data. You can use DDE to send faxes through WinFax, to send WinFax status information to other applications, and to enter WinFax commands from within other applications. Different Windows applications use different DDE command syntax, so it's not possible to provide universal DDE commands that will run WinFax from any application. But since WinFax only recognizes a limited number of commands, you'll need the information in this chapter (along with documentation supplied with the client application) to create WinFax macros.

If you're using Word for Windows, Ami Pro, or Excel, you can use the macros included in the WinFax package, rather than writing your own. Chapter 14 explains how to use these macros.

How DDE Works

A DDE conversation is an exchange of data between two application programs. DDE calls the application that initiates the conversation the *client*; the application that responds to the client is the *server*. The link between the client and the server is called the *channel*.

Every unit of DDE data has a three-part identity: *application, topic,* and *item*. The application is the server to which the data is directed. Topics are the type of action that the command performs. The item is the specific command or data in this transaction.

The application name is the filename of the executable program without the .EXE file extension. For both WinFax Lite and WinFax Pro, the application name is FAXMNG.

There are five kinds of DDE topics:

Initiate   The Initiate function opens a DDE channel. An Initiate command includes the server's application name and the topic. So a command to initiate a DDE conversation with WinFax might look like this:

DDEInitiate "FAXMNG" "Transmit"

When the server receives an Initiate command, it returns a channel number.

Request   A Request function instructs the server application to return data to the client. A Request command includes the channel number and an item.

Poke   A Poke function takes data from the client and sends it to the server. A Poke command includes the channel number, the item and the data that you want to "poke" to the server.

Execute   An Execute function sends commands from the client to the server. An Execute command includes the channel number and the execute string, which contains one or more commands.

Terminate   A Terminate function closes the DDE channel.

WinFax Lite Specifications

The application name for WinFax Lite is **FAXMNG**.

WinFax Lite supports only one topic, **Transmit,** which instructs WinFax to send a fax from within the client application. Transmit is a poke function.

You can poke two items to WinFax Lite: **Fax Number** and **Receiver**. The "Fax Number" is a character string that contains the telephone number that WinFax will dial; the "Receiver" is the recipient of the fax.

WinFax Pro Specifications

The application name for WinFax Pro is **FAXMNG**.

WinFax Pro supports two topics: **Transmit** and **Control**.

Transmit

Transmit is a poke function that sends data from the client application to WinFax. You can use the following items with the Transmit function.

Fax Number The telephone number to which WinFax will send the document.

Receiver The name of the recipient.

Sendfax This item is an instruction to transmit a fax to a specified recipient.

Sendfax uses these commands:

recipient

recipient(Fax Number, [Time], [Date], [Name], [Company], [Description], [Keywords], [Billing Code])

Recipient supplies the information that you would otherwise type into the Send Fax and Send Log dialog boxes. The recipient command must include the Fax Number variable; all the others are optional. Place the text of each variable inside quotation marks. If you want to skip a variable, include all the commas up to the variable you want to include. If you don't specify a time or date, WinFax will send the fax immediately. For example, to specify Mary Hamilton at 555-9734 as the destination of the current document, use this command:

recipient("555-9734",,,"Mary Hamilton")

Time to recharge my batteries!

showsendscreen

showsendscreen("1") | ("2")

Showsendscreen specifies whether the Send Fax dialog box appears when the macro prints a document to WinFax. Type **1** to display the Send Fax dialog box, or **2** to suppress it. The default is **2**.

resolution

resolution("HIGH") | ("LOW")

Resolution specifies the resolution of the next fax.

Control Control is a request topic that instructs WinFax to return status information to the client application.

WinFax Pro uses these Control commands:

Item DoneReceivingFax

Operation WM_DDE_ADVISE

This command from the client instructs WinFax to send a WM-DDE-DATA message back to the client when WinFax successfully receives a fax.

Item DoneReceivingFax

Operation WM_DDE_DATA

This is the message WinFax returns to the client when it receives a fax.

Item DoneReceivingFax

Operation WM_DDE_UNADVISE

This is an instruction from the client to the WinFax application to cancel a previous WM-DDE-ADVISE command.

Item NumberFaxesReceived

Operation WM_DDE_REQUEST

This is a request to WinFax for the number of received faxes currently in the WinFax event list.

Item TimeUntilNextOutgoing

Operation WM_DDE_REQUEST

This is a request to WinFax for the time of the next scheduled fax transmission.

Item   Status

Operation   WM_DDE_REQUEST

This is a request for the current WinFax status. WinFax returns one of these status messages:

BUSY	WinFax is sending or receiving a fax.
IDLE	WinFax is not sending or receiving, and automatic reception is not active.
ACTIVE	WinFax is not sending or receiving, and automatic reception is active.
REQUEST_ ACTIVE	WinFax has received a GoIdle command from the client, but a scheduled fax has not been sent.

Item   GoIdle

Operation   WM_DDE_EXECUTE

This is an instruction from an application that uses the same communications port as WinFax, such as a data communications program. When WinFax receives this command, it closes the port and turns off automatic reception.

Item   GoActive

Operation   WM_DDE_EXECUTE

This cancels the GoIdle command.

Item   ReceiveFaxNow

Operation   WM_DDE_EXECUTE

This is an instruction to perform a manual receive. Use this command when automatic reception is not active, with another program that might recognize an incoming call as a fax call.

APPENDIX

B

WinFax Lite
Variations

DELRINA has released several slightly different versions of Win-Fax Lite. They all share the same features, but they're organized slightly differently. Some versions are customized for a particular type of modem, while others contain minor changes to the WinFax Administrator screen.

The only way to obtain WinFax Lite is to get it in a package with a fax modem. Since Delrina wants you to buy a copy of WinFax Pro, they don't offer a WinFax Lite upgrade path. If your copy of WinFax Lite includes an Administrator screen that doesn't match the one described in Part II of this book, don't panic. You should still be able to find all of the commands in Chapter 7, but you might have to look in different menus. In this appendix, we'll explain the differences between the versions and show you where all the commands are hidden in both menu sets.

You can identify your WinFax Lite version by selecting the About command in the Help menu to display the About Delrina WinFax Lite dialog box shown in Figure B.1. Your copy may have a different date from the ones in this appendix, but the menus should match one of the versions described here.

Both versions have the same Help and Upgrade menus, but the other menus are slightly different. The version described in the body of this book has menus headed Fax, Send, Receive, and Phonebook, and only three commands in the button bar. As Figure B.2 shows, other versions have menus headed File, Send, Receive, and Phonebook. This version includes a button bar with eight commands.

Figure B.3 shows the individual menus in the May 20, 1993 version of WinFax Lite.

Figure B.4 shows the version of WinFax Lite described in Chapter 7 of this book. It has all the same commands as the other releases, but in a different arrangement.

FIGURE B.1

About WinFax dialog box, May 20, 1993 version

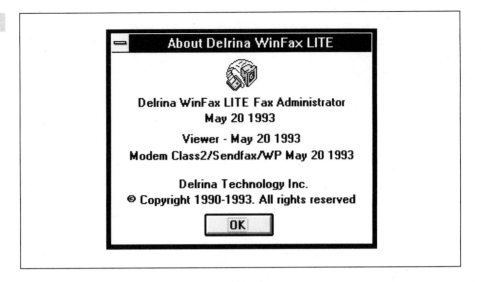

FIGURE B.2

Delrina WinFax Lite Administrator screen, May 20, 1993 version

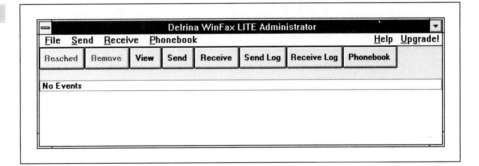

FIGURE B.3

WinFax Lite Administrator menus, May 20, 1993 version

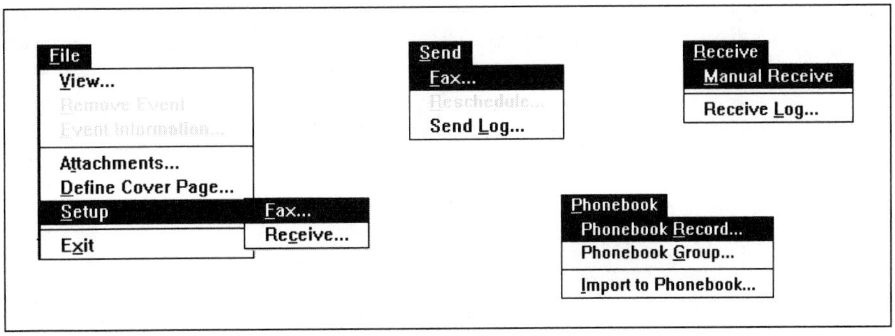

FIGURE B.4

The WinFax Lite
Administrator menus,
June 17, 1993 version

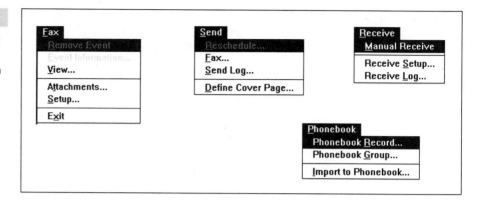

Even though the menus are different, you can find all the same commands in both versions of WinFax Lite if you know where to look. If you can't find a command as you work your way through this book, look in another menu. It's really more important to understand how to use a command after you find it, and the information in Part II applies equally to all versions.

Sorry I misse
your call.
I was having
an out-of-offic
experience.

INDEX

Note to the reader: Boldfaced page numbers indicate a primary reference to a topic. *Italicized* page numbers indicate a figure.

X

Z

GET A FREE CATALOG JUST FOR EXPRESSING YOUR OPINION.

Help us improve our books and get a *FREE* full-color catalog in the bargain. Please complete this form, pull out this page and send it in today. The address is on the reverse side.

Name _____ **Company** _____

Address _____ **City** _____ **State** ___ **Zip** _____

Phone (___) _____

1. How would you rate the overall quality of this book?

❑ Excellent
❑ Very Good
❑ Good
❑ Fair
❑ Below Average
❑ Poor

2. What were the things you liked most about the book? (Check all that apply)

❑ Pace
❑ Format
❑ Writing Style
❑ Examples
❑ Table of Contents
❑ Index
❑ Price
❑ Illustrations
❑ Type Style
❑ Cover
❑ Depth of Coverage
❑ Fast Track Notes

3. What were the things you liked *least* about the book? (Check all that apply)

❑ Pace
❑ Format
❑ Writing Style
❑ Examples
❑ Table of Contents
❑ Index
❑ Price
❑ Illustrations
❑ Type Style
❑ Cover
❑ Depth of Coverage
❑ Fast Track Notes

4. Where did you buy this book?

❑ Bookstore chain
❑ Small independent bookstore
❑ Computer store
❑ Wholesale club
❑ College bookstore
❑ Technical bookstore
❑ Other _____

5. How did you decide to buy this particular book?

❑ Recommended by friend
❑ Recommended by store personnel
❑ Author's reputation
❑ Sybex's reputation
❑ Read book review in _____
❑ Other _____

6. How did you pay for this book?

❑ Used own funds
❑ Reimbursed by company
❑ Received book as a gift

7. What is your level of experience with the subject covered in this book?

❑ Beginner
❑ Intermediate
❑ Advanced

8. How long have you been using a computer?

years _____

months _____

9. Where do you most often use your computer?

❑ Home
❑ Work

❑ Both
❑ Other _____

10. What kind of computer equipment do you have? (Check all that apply)

❑ PC Compatible Desktop Computer
❑ PC Compatible Laptop Computer
❑ Apple/Mac Computer
❑ Apple/Mac Laptop Computer
❑ CD ROM
❑ Fax Modem
❑ Data Modem
❑ Scanner
❑ Sound Card
❑ Other _____

11. What other kinds of software packages do you ordinarily use?

❑ Accounting
❑ Databases
❑ Networks
❑ Apple/Mac
❑ Desktop Publishing
❑ Spreadsheets
❑ CAD
❑ Games
❑ Word Processing
❑ Communications
❑ Money Management
❑ Other _____

12. What operating systems do you ordinarily use?

❑ DOS
❑ OS/2
❑ Windows
❑ Apple/Mac
❑ Windows NT
❑ Other _____

13. On what computer-related subject(s) would you like to see more books?

_____

_____

_____

_____

14. Do you have any other comments about this book? (Please feel free to use a separate piece of paper if you need more room)

_____

_____

_____

_____

PLEASE FOLD, SEAL, AND MAIL TO SYBEX

SYBEX INC.
Department M
2021 Challenger Drive
Alameda, CA
94501

SYBEX®

WinFax Lite Command Summary

All Version of WinFax lite have the same commands, but in different menus. If your Administrator screen has eight buttons, you have a Version A. If it has only two or three buttons, it's Version B.

VERSION A	VERSION B	COMMAND	ACTION
File	Fax	View...	Start the WinFax Image Viewer.
File	Fax	Remove Event	Delete the highlighted event.
File	Fax	Event Information...	Display details about the highlighted event.
File	Fax	Attachments...	Display a list of attachment files.
File	Send	Define Cover Page...	Create a cover page template.
File	Fax	Setup Fax...	Open the Setup dialog box to specify options.
File	Receive	Setup Receive...	Specify receive options.
File	Fax	Exit	Close the WinFax program.
Send	Send	Fax...	Send a fax from the WinFax Administrator.
Send	Send	Reschedule...	Change the transmission time or date of a previously scheduled fax.
Send	Send	Send Log...	Display a chronological list of previously transmitted faxes and failed attempts.
Receive	Receive	Manual Receive	Answer a ringing telephone line and accept an incoming fax.
Receive	Receive	Receive Log...	Display a chronological list of faxes that WinFax has received and failed attempts.
Phonebook	Phonebook	Phonebook Record...	Create a directory of names and fax numbers, or add to an existing directory.
Phonebook	Phonebook	Phonebook Group...	Create a new distribution list or edit an existing phonebook group.
Phonebook	Phonebook	Import to Phonebook...	Copy an existing ASCII text file into a phonebook, or create a new phonebook.
Help	Help	Help Index F1	Display Help for WinFax Lite.
Help	Help	About	Display copyright and release information.
Upgrade!	Upgrade!	WinFax PRO...	Order an upgrade to WinFax Pro. This menu disappears if you order the upgrade.